STUDIES IN GRAPH THEORY, PART II

Studies in Mathematics

The Mathematical Association of America

Studies in Mathematics

Volume 12

STUDIES IN GRAPH THEORY, PART II

D. R. Fulkerson, editor
Cornell University

Published and distributed by
The Mathematical Association of America

© *1975 by*
The Mathematical Association of America (Incorporated)
Library of Congress Catalog Card Number 75-24987

Complete Set ISBN 0-88385-100-8
Vol. 12 0-88385-112-1

Printed in the United States of America

Current printing (last digit):

10 9 8 7 6 5 4 3 2 1

ACKNOWLEDGMENTS

Professor Dantzig's contribution, "On the shortest route through a network", is reprinted, with slight changes, from MANAGEMENT SCIENCE, Volume 6, Number 2, January 1960; Professor Duffin's contribution, "Electrical network models", is an extension of his paper, "Network models", which appeared in SIAM-AMS PROCEEDINGS, Volume 3 (1971), pages 65–91; Professor Fulkerson's contribution, "Flow networks and combinatorial operations research", is reprinted, with slight changes, from the AMERICAN MATHEMATICAL MONTHLY, Volume 73, Number 2, February 1966; Dr. Gomory's and Professor Hu's contribution, "Multi-terminal flows in a network", is reprinted from the SIAM JOURNAL ON APPLIED MATHEMATICS, Volume 9, Number 4, December 1961; Professor Minty's contribution, "On the axiomatic foundations of the theories of directed linear graphs, electrical networks and network-programming", is reprinted, with slight changes, from the JOURNAL OF MATHEMATICS AND MECHANICS, Volume 15, Number 3 (1966); Professor Whitney's and Professor Tutte's contribution, "Kempe chains and the four colour problem", is reprinted from UTILITAS MATHEMATICA, Volume 2, November 1972.

PREFACE

It is probably fair to say, and has been said before by many others, that graph theory began with Euler's solution in 1735 of the class of problems suggested to him by the Königsberg bridge puzzle. But had it not started with Euler, it would have started with Kirchhoff in 1847, who was motivated by the study of electrical networks; had it not started with Kirchhoff, it would have started with Cayley in 1857, who was motivated by certain applications to organic chemistry, or perhaps it would have started earlier with the four-color map problem, which was posed to De Morgan by Guthrie around 1850. And had it not started with any of the individuals named above, it would almost surely have started with someone else, at some other time. For one has only to look around to see "real-world graphs" in abundance, either in nature (trees, for example) or in the works of man (transportation networks, for example). Surely someone at some time would have passed from some real-world object, situation, or problem to the abstraction we call graphs, and graph theory would have been born.

Today graph theory is a vast and somewhat sprawling subject, embracing as it does applications in many diverse areas: physics, chemistry, engineering, operations research, genetics, economics, psychology, and sociology, to name some. Dozens of books and proceedings of conferences on graph theory have appeared, mostly within the last fifteen years, and the number of journal articles dealing with graphs that have appeared in this time interval must number in the thousands. Today there are journals devoted exclusively to graph or network theory, and other journals, devoted exclusively to combinatorial mathematics, in which many, if not most, of the papers that appear are about graphs.

This recent explosion in a subject that was fairly dormant over a long period of time creates a difficult situation for one who is asked to edit a study on graph theory. Many facets of the subject must be omitted entirely; others can be treated in only a sketchy fashion. The resulting study will be biased by the editor's ignorance of some topics in the subject, and by his likes and dislikes for topics he knows something about. These remarks would apply to almost any editor; they certainly apply to me. Some of the important omissions that I know about include the fairly recent and lengthy affirmative resolution of the Heawood map conjecture by Ringel and Youngs, the solution of the Shannon switching game by Lehman, and the work of Edmonds on weighted matching theory, together with its application to very practical generalizations of the Euler problem. The latter would have brought us back to where it all started.

I shall let the papers that comprise the two volumes of this study speak for themselves. Some of them have appeared elsewhere; others appear here for the first time.

D. R. FULKERSON

CONTENTS

POLYTOPAL GRAPHS*

Branko Grünbaum

A graph G is called *d-polytopal* provided there exists a d-dimensional convex polytope P such that the vertices and edges of G are in a one-to-one incidence-preserving correspondence with those of P. In other words, G is d-polytopal if and only if it is isomorphic to the 1-skeleton of some convex d-polytope P. The polytope P is then said to *realize* G, and G is called *the graph* of P. Instead of "3-polytopal graph," we shall often say "polyhedral graph."

Convex polygons and polyhedra (that is, 3-dimensional convex polytopes) have been frequent topics of investigation since antiquity; about a century ago, their study was extended to polytopes of arbitrary dimension d. While much of the early interest was of a metric character (Pythagorean theorem, regular polygons, Platonic and Archimedean solids, etc.), in more recent times the combinatorial point of view has attracted most of the attention. Among problems of this type, two deserve particular mention as they had great influence on the study of polytopal graphs.

*Research supported in part by the Office of Naval Research under Grant N00014-67-A-0103-0003.

Two polytopes are called *isomorphic* (sometimes also "combinatorially equivalent") provided their faces of all dimensions (vertices, edges, . . . , facets) can be brought into a one-to-one inclusion-preserving correspondence. The determination of all *isomorphism types* of polygons is trivial: for every $n \geqslant 3$, there exist *n-gons* (that is, convex polygons with n sides) and every two *n*-gons are isomorphic. But already in dimension 3 the problem is very difficult. Leonhard Euler, Jacob Steiner, Arthur Cayley, and others, worked on the determination of the number of non-isomorphic polyhedra with a given number v of vertices (or of suitable subclasses of such polyhedra) with no success beyond the experimental solution for very small values of v. (It should be mentioned that Euler's study of the question led him to the discovery of the "Euler formula" relating the numbers of vertices, edges and faces of a polyhedron, which was one of the starting points of modern topology.) As we shall see below, the isomorphism of polyhedra is equivalent to the isomorphism of their graphs; thus an important geometric problem becomes translated into a problem in graph theory. The reader interested to learn about the present state of the enumeration problem for polyhedra and about references to its history and literature should consult Federico [1], [2] and Tutte [4].

The second problem which has exerted great influence on the development of the theory of polytopal graphs was the question of efficiency of linear programming and other computational techniques. It led to a renewal of interest in the combinatorial theory of polytopes in general, and to various questions regarding paths in polytopal graphs in particular, thus motivating many of the investigations. An account of the relevant results and their history, with numerous references to the literature, may be found in the stimulating paper of V. Klee [2].

Polytopal graphs are endowed with many remarkable properties; their study naturally leads to a large number of questions in graph theory, combinatorics, topology and geometry. A few such questions will be mentioned later.

The following discussion is meant to impart to the reader some feeling about the type of results known on polytopal graphs, the methods used in proving them, and the open questions related to them. We attempted to provide the reader with a few historical facts in the hope that he will share our enjoyment of the interplay between old, almost forgotten knowledge and new ideas and points of view. Without striving for completeness, we have also attempted to include enough bibliographic references to enable the interested reader to locate the original publications.

Most of the material deals with 3-polytopal graphs. This is natural since their properties have been investigated in more detail, and are easier to understand without detailed technical background, than those of d-polytopal graphs with $d \geqslant 4$. Some aspects of the latter are discussed in the last part of the report.

One of the simplest and most elegant general results on polytopal graphs was first established by M. L. Balinski [1] in 1961:

THEOREM 1. *Every d-polytopal graph is d-connected.*

In other words, any two vertices of a d-polytopal graph are connected by d paths that are pairwise disjoint except for their endpoints.

Theorem 1 clearly supplies a partial answer to the fundamental problem of characterizing, for each d, the family of d-polytopal graphs. It is obvious that for $d = 2$ this characterization is trivial: G is 2-polytopal if and only if G is a circuit with $n \geqslant 3$ vertices and edges. But the following characterization of 3-polytopal graphs found by E. Steinitz [1] in 1916 is one of the most remarkable results about polytopal graphs:

THEOREM 2: *A graph G is polyhedral if and only if G is planar and 3-connected.*

Easily accessible proofs of Steinitz' theorem may be found in Grünbaum [1] and in Barnette-Grünbaum [1]. A proof that parallels part of one of Steinitz's original proofs is attempted in Lyusternik [1]; unfortunately, Lyusternik's formulation (in the Russian original, as well as in the two English translations) is fallacious in failing even to mention the need for an argument at the crucial stage of one of the main steps of the proof, and also in failing to include conditions that would force the graph to be 3-connected.

With Theorem 2, we arrived at one of the reasons for the importance of polytopal graphs: like the 4-color problem, many of the interesting questions about planar graphs deal with 3-connected graphs, or can easily be reduced to deal with such graphs. But then they become questions about polyhedral graphs, and the scene is set for mutual influences of geometric and combinatorial ideas.

Before pursuing this direction in some detail, it appears worthwhile to make a few comments on the history, proof, and meaning of Theorem 2.

Steinitz was aware of the basic nature of his result; he called it the "Fundamental Theorem on Convex Types [of Polyhedra]" and —in analogy to Gauss' four proofs of the "Fundamental Theorem of Algebra"—gave three completely different proofs. The details are meticulously worked out in the book by Steinitz and Rademacher [1]. It should be remarked, however, that Steinitz established his theorem before the emergence of graph theory and the notions of connectivity and planarity. Hence he was obliged to use other notions and a different terminology; he works with 2-dimensional complexes, and so his formulation of the result as well as the details of the proofs are rather cumbersome.

The assertion that every polyhedral graph G is 3-connected is clearly just a special case of Theorem 1; it may also be easily established in a direct way. The planarity of each polyhedral graph G may be proved using the so-called *Schlegel diagrams* (named after V. Schlegel [1] who introduced them in 1881 in connection with his investigations of 4-dimensional regular polytopes). Let P be a polyhedron that realizes G; imagining P made of cardboard, we omit one of its faces and view the opening from a point sufficiently close by to see all the other faces of P from the inside. A *Schlegel diagram* of P is any projection of all the vertices and edges of P from such a point into the deleted face; thus each Schlegel diagram appears as a partition (tessellation) of a convex polygon into convex polygons. The existence of Schlegel diagrams

for each polyhedron P establishes the planarity of every polyhedral graph G.

The other half of Steinitz' theorem is the difficult one. It
amounts to the construction of a polyhedron P that realizes a
given 3-connected planar graph G. The reader will possibly have
an inkling of some of the difficulties involved by studying the
various parts of Figure 1. All the graphs in it are mutually
isomorphic, and although people endowed with a good 3-
dimensional intuition may imagine a convex solid realizing that
graph G, it is rather hard to describe the construction of the solid
from the graph. (The reader may wish to convince himself that
none of the shown representations of G is a Schlegel diagram of
any convex polyhedron.) It should also be noted that the graph G
in Figure 1 has several features that make the construction of a
solid realizing it uncharacteristically easy.

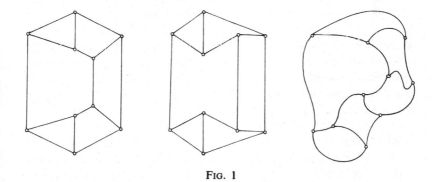

FIG. 1

All the known proofs of Steinitz' theorem proceed by induction
on the number of edges of G, the assertion being obviously true
for the unique 3-connected planar graph G with 6 edges (as well as
for some other classes of graphs). In the inductive step there are
two stages: (i) By a suitable method, to each 3-connected planar
graph G with more than 6 edges a graph G' is associated that is of
the same type but has fewer edges than G. (ii) A method is
described by which from each polyhedron P' that realizes G' a

polytope P realizing G may be constructed. The different proofs vary in the methods used in the two stages; for details, the reader should consult Steinitz-Rademacher [1], Grünbaum [1] or Barnette-Grünbaum [1].

It is rather curious to observe that many mathematicians seemed (and still seem) to have an intuitive feeling that any planar graph that somehow resembles a Schlegel diagram is polyhedral, and that a polyhedron realizing it may be obtained by "lifting" part of the graph above the plane. But more curious still is the fact that many—Schlegel among them—did not seem to feel that this is an assertion needing some proof. One of the most interesting of such cases is that of T. P. Kirkman. In 1857, Kirkman discovered the planar variant of the characterization of 3-connected graphs obtained by Tutte [3] in 1961—but Kirkman was not aware of the need of justifying the possibility of moving various vertices, that possibility being completely obvious for planar graphs but not at all evident for vertices of polyhedra. The importance of Steinitz's theorem is, among others, that it provides post-factum justification for many such arguments. (By the way, much of the work of the Reverend Thomas P. Kirkman is undeservedly forgotten. He was an original and industrious investigator in geometry and combinatorics, and some of his results will be mentioned later. Kirkman antedated Steiner in posing, and Reiss in solving, the problem of "Steiner triples," invented "Hamiltonian" circuits, etc.)

Among the many ramifications of Steinitz's theorem, we mention only the following few:

(2.a) Every maximal planar graph (that is, triangulation of the plane or the sphere) with at least 4 vertices is polyhedral.

(2.b) The faces ("countries") of every 3-connected planar graph are uniquely determined by the graph. (Whitney's theorem.)

(2.c) A 3-connected planar graph G and its dual graph G' may be realized by polyhedra P and P' that are polar (or dual) to each other. (In this context, it should be mentioned that the treatment of dual graphs in many texts is obviously patterned on 3-connected graphs although the statements are frequently formulated in a generality in which they are no longer valid.)

(2.d) Every planar graph has planar realizations in which each edge is a rectilinear segment. (Wagner's theorem.)

(2.e) Every 3-connected planar graph has realizations in the plane in which all the bounded countries are convex, as is the complement of the unbounded country.

(2.f) Every polyhedral graph G has a realization by a polyhedron P such that all automorphisms of G are induced by symmetries of P. (Mani [1].)

For details about these results and their literature, see Grünbaum [2].

As a challenging open problem related to (2.a), we mention the following (compare Duke [1] for other problems, known results, and references): If a graph G triangulates the torus, does there exist an isomorphic triangulation of a torus such that each of its triangles is a geometric (rectilinear) triangle? An open problem related to (2.e) is due to Ungar [1]: Does there exist a constant $c > 0$, such that for each 3-valent polyhedral graph G there is an imbedding of G in the plane with all faces convex, and for each face the ratio of its incircle to its circumcircle (or the ratio of its width to its diameter) is at least c? Also unsolved is the question: Which imbeddings of a 3-connected planar graph are actually obtainable as Schlegel diagrams of suitable polyhedra?

We turn now to a discussion of several other properties of 3-polytopal graphs.

One line of investigations, motivated at least in part by the problem of enumeration of polyhedra, was substantially advanced by V. Eberhard [1] in 1891. To simplify the exposition, let us agree to denote by v, e, p the numbers of vertices, edges, and faces of a polyhedron or of a polyhedral graph, and by v_k and p_k the numbers of k-valent vertices and k-sided faces. If we wish to indicate which polyhedron P or graph G is considered, we shall write $v(P), v(G)$, etc. With those symbols Euler's relation reads $v - e + p = 2$, and it is easily seen that we also have the following relations:

$$v = \sum_{k \geqslant 3} v_k, \qquad p = \sum_{k \geqslant 3} p_k,$$

$$2e = \sum_{k \geqslant 3} k v_k = \sum_{k \geqslant 3} k p_k.$$

Appropriate combinations yield, among others,

$$\sum_{k>3} (6 - k)p_k + 2 \sum_{k>3} (3 - k)v_k = 12, \qquad (*)$$

$$\sum_{k\geqslant 3} (4 - k)(p_k + v_k) = 8, \qquad (**)$$

from which it follows at once that $p_3 + p_4 + p_5 \geqslant 4$ and $p_3 + v_3 \geqslant 8$.

Eberhard was the first to consider questions of the following type: Given finite sequences (p_k) and (v_k) of non-negative integers that satisfy (*) or (**), does there exist a polyhedron or a 3-polytopal graph with those preassigned values of p_k and v_k? Eberhard's main result concerns the case of relation (*) provided $v_k = 0$ for all $k \geqslant 4$; he proved:

THEOREM 3: *For any finite sequence* (p_k) *of non-negative integers satisfying*

$$\sum_{k>3} (6 - k)p_k = 12, \qquad (***)$$

there exists a 3-valent polyhedral graph G *with* $p_k(G) = p_k$ *for all* $k \neq 6$.

Eberhard's original proof was based on a detailed investigation of systems of hexagons that form part of the boundary of a convex polyhedron and was very involved. Using Steinitz' theorem, it is possible to prove this result (and many others) by the much simpler construction of appropriate 3-connected planar graphs. The main idea in those proofs is to put together, in a suitable way, convenient standardized "building blocks," each of which accounts for one of the "large" polygons; in certain cases additional "repairs" must be performed. A detailed proof along those lines may be found in Grünbaum [1].

Eberhard's result served as a starting point for many different investigations, some of which we shall now briefly mention; their proofs are mostly based on suitable variations of the above idea.

(3.a) If sequences (p_k) and (v_k) satisfy (**) and if $\sum_{k\geqslant 3} kv_k$ is even, then there exists a polyhedral graph G with $p_k(G) = p_k$ and $v_k(G) = v_k$ for all $k \neq 4$.

(3.b) If in Theorem 3 we have $p_3 = p_4 = 0$, then even $p_6(G)$ may be required to have any preassigned value $\geqslant 8$. In any case, the "gaps" in the possible values of p_6 are not great, as follows from the recent result of Fisher [1]: If a sequence (p_k) satisfies (***), there exists a constant $m_0 \leqslant 3\Sigma_{k \neq 6} p_k$, such that for each choice of $p_6 = m_0 + 2m$, m a positive integer, there is a 3-valent polyhedral graph G with $p_k(G) = p_k$ for all k.

(3.c) There exist non-trivial lower bounds on $p_6(G)$ in terms of $p_k(G)$, $k \neq 6$, valid for all 3-valent polyhedral graphs G (Barnette [2]; Jucovič [1]); for example, Jucovič established

$$3p_6(G) \geqslant 12 - 2p_4(G) - 3p_5(G) + \sum_{k > 7} ([\tfrac{1}{2}(k + 1)] - 6)p_k(G).$$

(3.d) If 2-connected planar graphs G are considered, 2-valent vertices and digons may be reasonably admitted; their numbers may be denoted by $v_2(G)$ and $p_2(G)$. Similarly, if G is assumed only to be connected, then we may have $v_1(G) > 0$ and $p_1(G) > 0$. Theorem 3 and remark (3.a) have analogues in those instances as well, with 2-connected, or connected, planar graphs instead of polyhedral graphs (Rowland [1]).

(3.e) All the results mentioned above have analogues valid for graphs imbedded in orientable 2-manifolds other than the sphere (or the plane); the case corresponding to equation (*) for a manifold of genus g has been settled by Jendrol'-Jucovič [1], where the reader may also find references to the rather abundant literature.

(3.f) As conjectured already by Eberhard in 1891 but first proved only in 1964 by T. S. Motzkin, if a 3-valent polyhedral graph G is a multi-3-gon (that is, satisfies $p_k(G) = 0$ whenever k is not a multiple of 3) then $p(G)$ is even. Many extensions and variants of this result are known (see, in particular, Malkevitch [1], Gallai [1], Medyanik [1]), although the understanding of the phenomenon is still rather lacunary. Possibly this is to be expected in view of the fact that the 4-color problem is equivalent to the question whether every 3-valent polyhedral graph G can be made into a multi-3-gon by "cutting off" suitable vertices of G (that is, replacing each of them by a triangular face).

A challenging open problem in this area is the characterization of the sequences (v_k) for which there exist polyhedral graphs, or just planar graphs, G such that $v_k(G) = v_k$ for all k. This question was posed (in a dual formulation) already by Sainte-Marie [1] in 1895, but only very superficial partial answers are known (see Grünbaum [2, p. 1142] for references to the literature).

Related to (3.d) and (3.f) is a remarkable problem about 3-valent 2-connected planar graphs, such that $p_2 = 3$ and $p_k = 0$ for $k \neq 2,6$. It was conjectured by Brunel [1] in 1897 that a value of p_6 is possible for such graphs if and only if $p_6 = x^2 + xy + y^2 - 1$, where x and y are non-negative integers with $x^2 + y^2 > 0$. The problem was independently raised by Malkevitch [2] in 1970, but a solution has still not been found.

A different direction of investigations concerns "Hamiltonian" and other circuits and paths in polyhedral graphs. Introduced by T. P. Kirkman [1] in 1855, simple circuits through all the vertices of a polyhedral graph appear in surprisingly varied contexts. (We shall not attempt to correct the historical injustice by trying to substitute "Kirkmanian" for "Hamiltonian.") Kirkman used them in deriving a method of representing certain polyhedra; he observed that not every polyhedron admits a Hamiltonian circuit and gave an example of one that did not (the example is reproduced in Steinitz [1, p. 49]).

A famous conjecture proposed by Tait [1] in 1880 asserts that every 3-valent polyhedral graph admits a Hamiltonian circuit. A proof of Tait's conjecture would lead to an affirmative solution of the 4-color problem; it is also relevant to some questions of systematics of cyclic compounds in organic chemistry, as pointed out by J. Lederberg in 1966. Although already Kirkman [2] in 1881 doubted the validity of Tait's conjecture, the first counterexample to it was found only in 1946 by W. T. Tutte [1]. A brilliant—but post factum almost trivial—method of generating such graphs was found by Grinberg [1] in 1968; it is rather touching to realize that Kirkman in 1881 had essentially the same idea and condition—but due to notational clumsiness stopped just short of the goal. Kirkman's equation (F) on page 113 is, except for notation and trivial uses of the Euler relation, identical to Grinberg's condition: If a 3-valent polyhedral graph G admits a Hamiltonian circuit, then there exists a decomposition $p_k(G) = p'_k + p''_k$ (with non-negative integers p'_k, p''_k), such that

$$\sum_{k>3} (k - 2)p'_k = \sum_{k>3} (k - 2)p''_k.$$

Considering this equation modulo 3, it is immediate that the graph G, obtained from the graph G' in Figure 2 by "shrinking" each small triangle to a vertex, has no Hamiltonian circuit since $p_4(G) = 1$, $p_5(G) = 18$, $p_8(G) = 4$ and $p_j(G) = 0$ for $j \neq 4$, 5, 8. (Since in a 3-valent graph "cutting off" vertices does not affect the existence or non-existence of a Hamiltonian circuit, it follows that the graph G' shown in Figure 2 has no Hamiltonian circuit.)

Among the ramifications of the fact that polyhedral graphs—and even 3-valent ones—may fail to have Hamiltonian circuits we mention the following:

(4.a) A deep result of Tutte [2] establishes that every 4-connected polyhedral (that is, 4-connected planar) graph possesses Hamiltonian circuits. Concerning extensions to graphs imbedded in 2-manifolds see Duke [2].

(4.b) G. Ewald [1] recently established that every triangulation of the plane in which each vertex is of valence at most 6 has a Hamiltonian circuit; he also proved a number of related results on other families of polyhedral graphs. (The existence of Hamiltonian circuits in 6-valent triangulations of the torus was established by Altshuler [1].)

(4.c) For various classes of polyhedral graphs there exist constants $\alpha < 1$ and β (depending on the class) such that $h(G) \leqslant \beta(v(G))^\alpha$ for each G of the class, where $h(G)$ denotes the maximal length of a simple circuit in G. For a survey of results and literature on that topic and for improvements of many of the previous results, see Grünbaum-Walther [1]. For the class of all polyhedral graphs one may choose any $\alpha > \log 2/\log 3$ (Moon-Moser [1]); it may be conjectured that $\alpha \geqslant \log 2/\log 3$ for every infinite family of polyhedral graphs.

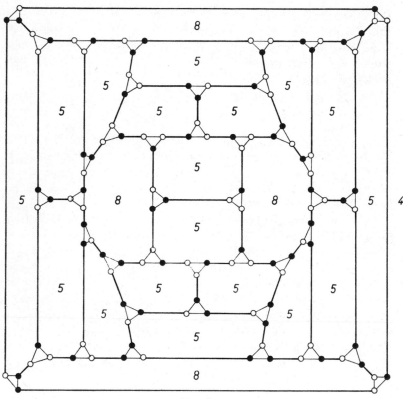

FIG. 2

(4.d) Let $\Gamma(j, m)$ denote the family of all graphs G, such that $v(G) - h(G) = m$ and that for every j vertices of G there exists a circuit of length $h(G)$ missing those j vertices; $\Gamma_0(j, m)$ denotes the subfamily of $\Gamma(j, m)$ consisting of polyhedral graphs. The graphs belonging to $\Gamma(1, 1)$ are usually called "hypohamiltonian;" they have recently been investigated by several authors, who also dealt with various related notions. It is known (see Grünbaum [5] for this result and for references to earlier literature) that $\Gamma_0(1, 3) \neq \emptyset$; it may be conjectured that $\Gamma_0(1, 1) = \Gamma_0(1, 2) = \emptyset$, and that each $\Gamma_0(j, m)$ contains only finitely many non-isomorphic graphs. On the other hand we conjecture that for each j there is an m_j such that $\Gamma_0(j, m_j) \neq \emptyset$.

Before turning our attention to d-polytopal graphs for $d \geqslant 4$, we would like to focus attention on a few additional results and problems dealing with polyhedral graphs.

A surprising result of Barnette [1] is:

THEOREM 5: *Every polyhedral graph G contains a spanning tree (that is, a tree containing all the vertices of G) of maximal valence 3.*

Among open questions related to Theorem 5 are:

If G is a polyhedral graph and G^* a graph dual to G, does there exist a spanning tree T of G with maximal valence 3, such that the edges of G not in T correspond in G^* to a spanning tree of G^* with maximal valence 3?

Does every 3-connected graph imbedded in the torus have a spanning tree of maximal valence 3? What are the analogues of Theorem 5 for 3-connected graphs imbeddable in an orientable (or else, in a non-orientable) manifold of genus g?

A beautiful result of Kotzig [1] deals with the *weight* $w(E)$ of an edge E of a graph G, where $w(E)$ is defined as the sum of the valences of the two vertices of E.

THEOREM 6: *Every polyhedral graph G contains an edge E such that $w(E) \leqslant 13$, and this inequality is best possible.*

Since Kotzig's result is not well known and its proof is not readily accessible, we bring here an outline of the arguments. The assertion that the result is best possible is established by the graph in Figure 3. For the proof of the inequality, we first note that if a polyhedral graph G_1 is not a triangulation, but has the property that $w(E) \geqslant k$ for each edge E of G_1, then by adding suitable diagonals G_1 may be enlarged to a triangulation G_2 of the plane such that $w(E) \geqslant k$ for each edge E of G_2. Thus it is enough to prove the inequality of the theorem for triangulations G. Let $e_{j,k}$

denote the number of edges of G that have one vertex of valence j, the other of valence k. Assuming that $e_{j,\,k} = 0$ for all pairs (j, k) such that $j + k \leqslant 12$, we consider the $3v_3$ edges incident with the v_3 vertices of valence 3. Counting from the other endpoints of those edges, and observing that two adjacent edges may not both lead to vertices of valence 3, we obtain

$$3v_3 \leqslant e_{3,\,10} + \sum_{k\,\geqslant\,11} [\tfrac{1}{2}k]v_k.$$

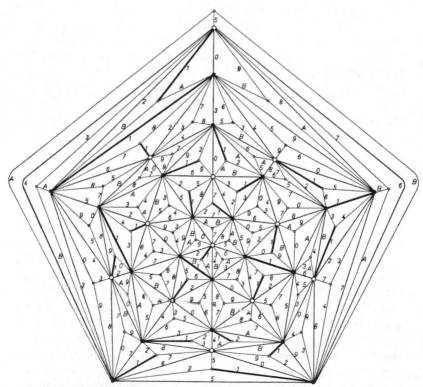

FIG. 3. A polyhedral graph with 12 edge-disjoint matchings. Each matching may be mapped onto each other by a self-isomorphism of the graph. One matching is emphasized by heavy edges.

Similarly, considering the $3v_3 + 4v_4$ edges incident with the vertices of valence at most 4, or the $3v_3 + 4v_4 + 5v_5$ edges incident with vertices of valence at most 5, we obtain

$$3v_3 + 4v_4 \leqslant \sum_{k > 9} [\tfrac{1}{2}k]v_k,$$

$$3v_3 + 4v_4 + 5v_5 \leqslant \sum_{k > 8} [\tfrac{1}{2}k]v_k.$$

Multiplying these inequalities by 5, 3, and 2 and adding, we obtain

$$30v_3 + 20v_4 + 10v_5$$

$$\leqslant 5e_{3,\,10} + 8v_8 + 20v_9 + 25v_{10} + 10 \sum_{k \geqslant 11} [\tfrac{1}{2}k]v_k.$$

But by Euler's relation $3v_3 + 2v_4 + v_5 = 12 + \sum_{k \geqslant 7} (k - 6)v_k$, so that

$$120 + 10v_7 + 12v_8 + 10v_9 + 15v_{10} + 10 \sum_{k \geqslant 11} ([\tfrac{1}{2}(k + 1)] - 6)v_k$$

$$\leqslant 5e_{3,\,10}.$$

Therefore, $e_{3,\,10} \geqslant 24$, thus establishing Theorem 6.

Among applications of Theorem 6 and of the related results about polyhedral graphs with $v_3 = 0$ and/or with $v_4 = 0$, we mention the following:

(6.a) A set M of edges of a graph G is called a (set-wise maximal) *matching* of G provided the edges in M are disjoint but each edge of G meets some edge in M. Then Theorem 6 implies that the maximal number of edge-disjoint matchings of a polyhedral graph is 12; in the graph of Figure 3 the numerals 0, 1, . . . , 9 and the letters A, B indicate a family of 12 edge-disjoint matchings. (See Grünbaum [6] for this and other results on matchings.) It may be conjectured that in every polyhedral graph there exist at least 3 edge-disjoint matchings. Another open problem is whether every polyhedral graph G contains a matching with at most $2v(G)/5$ edges.

(6.b) The vertices of every planar graph G may be colored by 5 colors α, β, γ, δ, ϵ in such a manner that no edge connects vertices of the same color and each of the graphs $G(\alpha, \beta)$, $G(\gamma, \delta)$, $G(\gamma, \epsilon)$ and $G(\delta, \epsilon)$ is acyclic, where $G(\lambda, \mu)$ is the subgraph of G spanned by the vertices of colors λ and μ. (See Grünbaum [4] for a proof of this and some related results, and for references to previously known results in this area.) It is not known whether the vertices of each planar graph G may be colored by 5 colors, α, β, γ, δ, and ϵ, so that all the graphs $G(\beta, \gamma)$, $G(\beta, \delta)$, $G(\beta, \epsilon)$, $G(\gamma, \delta)$, $G(\gamma, \epsilon)$ and $G(\delta, \epsilon)$ are acyclic.

The graph G of the polyhedron obtained from the dode-cahedron by placing a 5-sided pyramid on each of its 12 faces is an example of a 5-connected polyhedral graph that is cyclically-11-connected. (A graph G is *cyclically-k-connected* if the removal of less than k edges cannot disconnect it into two components, each of which contains a circuit.) Although there exist polyhedral graphs (and even 4-connected ones) of arbitrarily high cyclic connectivity, a remarkable result of Plummer [1] establishes that no 5-connected polyhedral graph can be cyclically-14-connected. It may be conjectured that no 5-connected polyhedral graph can by cyclically-12-connected, and that either Kotzig's theorem or the ideas used in its proof may lead to a proof of this conjecture.

Many results are known concerning d-polytopal graphs with $d \geqslant 4$, but for the main problem no solution is in sight. That problem is the characterization of d-polytopal graphs for each d. The difficulties seem to have several sources, at least two of which can be readily identified:

First, in contrast to the situation for $d \leqslant 3$, in case $d \geqslant 4$ the condition of d-polytopality does not force the graph to have relatively few edges. Indeed, for $d \geqslant 4$ the complete graph K_n with n vertices is d-polytopal whenever $n \geqslant d + 1$.

Second, for $d \geqslant 4$ a d-polytopal graph may happen to be realizable by d-polytopes which are not isomorphic to each other, and even by polytopes of different dimensions. Thus not only do we have no analogue of Steinitz's theorem for $d \geqslant 4$, but even Whitney's theorem (see (2.b) above) does not generalize. (A similar ambiguity of generalizations complicated the extension of the well-known planarity condition of Kuratowski to the question of imbeddability in manifolds of higher genus.)

One of the promising directions of investigation is to inquire about properties of higher-dimensional skeleta of polytopes— although then difficulties of another nature are frequently encountered. It would lead us too far to try to survey here the various results known about graphs and higher-dimensional skeleta of d-polytopes for $d \geqslant 4$. Many of them were established or surveyed in Grünbaum [1], [2]; in the present paper, we shall recall only two such results on polytopal graphs, and then report on some newer investigations.

As a relative of Theorem 1 we have:

THEOREM 7: *Every d-polytopal graph, $d \geqslant 2$, contains a subdivision of the graph K_{d+1}.*

It is of some interest to note that the simplest proof known for Theorem 7 actually establishes the corresponding result for polytopal skeleta of all dimensions.

By way of comparing Theorems 1 and 7, and as an example of influence of polytopal graphs on geometry, we mention the result of Larman-Rogers [1] that meaningfully generalizes Theorem 1 to the 1-skeleton of every d-dimensional compact convex set; the corresponding generalization of Theorem 7 is still open.

In order to establish (for each d) the existence of d-polytopal graphs which are dimensionally unambiguous (that is, graphs that are not e-polytopal for $e \neq d$) Klee [1] introduced the notion of "degree of total separability" of a graph G. We shall say that a set A of vertices of a graph G is *totally separated* by a set B of vertices provided A and B are disjoint and every path in G with endpoints in A meets B. The nth degree of total separability $s_n(G)$ of G is the largest cardinality of a set of vertices of G totally separated by some set of n vertices of G. Klee's surprising result is:

THEOREM 8: *The maximum of values of $s_n(G)$ possible for d-polytopal graphs G with $d + 1 \leqslant n$ equals $\mu(n, d)$, the maximal* *number of facets* (*that is,* $(d-1)$-*faces*) *possible for a d-polytope with n vertices.*

The values of $\mu(n, d)$ (which are relevant also to linear programming) formed the topic of many investigations. Extending previously known results, McMullen [1] established the relation

$$\mu(n, d) = \binom{n - [(d+1)/2]}{n - d} + \binom{n - [(d+2)/2]}{n - d}$$

for all $n \geqslant d + 1$.

The function $s_n(G)$ will probably be useful in other graph-theoretic investigations as well. Part of the proof of Theorem (9.a) below relies on it. On the other hand, Chvátal [1] found that $t(G) = \min_n \{n/s_n(G)\}$ is of interest in the study of Hamiltonian circuits. Among his results is the following:

(8.a) If G is a planar graph with $t(G) > 3/2$, then G has a Hamiltonian circuit.

This is probably the best possible result; the 63 dark vertices of the non-Hamiltonian graph G' in Figure 2 show that $t(G') \leqslant 3/2$, and it may be conjectured that $t(G') = 3/2$.

Among the open problems related to the facts just mentioned are:

Do there exist for each $d \geqslant 4$ d-polytopal graphs that are *completely unambiguous*? (A graph G is completely unambiguous provided every two polytopes that realize G are isomorphic as polytopes by the correspondence of their vertices which establishes them as realizations of G.) The existence of such graphs is known for $d \leqslant 5$ (see Barnette [3]) but the question is open for all larger values of d.

If a graph is d'-polytopal as well as d''-polytopal for some $d'' > d'$, is it necessarily d-polytopal for all d such that $d' \leqslant d \leqslant d''$?

A few results are known about matchings of polytopal graphs G. We denote by $\underline{m}(G)$ and $\overline{m}(G)$ the least and the largest numbers of edges in a matching of G, and by $\underline{g}(v, d)$ and $\overline{g}(v, d)$ the minima of $\underline{m}(G)$ and $\overline{m}(G)$ for d-polytopal graphs G with v

vertices. Clearly $\underline{g}(v, 2) = [(v + 2)/3]$ and $\bar{g}(v, 2) = [v/2]$. The following results are among those proved in Grünbaum [6]:

(9.a) There exist positive constants c'_d and c''_d, $d \geqslant 2$, such that

$$c'_d v^{1/[d/2]} \leqslant \underline{g}(v, d) \leqslant \bar{g}(v, d) \leqslant c''_d v^{1/[d/2]}.$$

(9.b) $\underline{g}(v, 3) = [(v + 9)/6]$.

The values of $\bar{g}(v, 3)$ seem not to have been determined (see also (6.a) above).

R. Forcade [1] recently established the following unexpected result, in which I_d denotes the graph of the d-dimensional cube (we recall that $v(I_d) = 2^d$.).

(9.c) $\lim_{d \to \infty} \underline{m}(I_d)/v(I_d) = 1/3$.

Results similar in spirit to the above probably hold for *covers* of d-polytopal graphs G, that is sets of edges that contain all vertices of G but have no proper subsets with that property.

As a last topic we consider the *cut-numbers*, which are actually concerned not with polytopal graphs but with their realizations by vertices and edges of polytopes. If P is a d-polytope, a *cut* of P is any set of edges of P that may be simultaneously intersected by a $(d - 1)$-dimensional hyperplane that misses all the vertices of P. We denote by $c(P)$ the maximal number of edges in a cut of P, and we define the *cut-number* $k(P)$ of P as the minimal number of cuts needed to cover all the edges of P. Clearly $k(P)c(P) \geqslant e(P)$, the number of edges of P.

A recent result of O'Neil [1] deals with cuts of the d-dimensional (regular) cube I^d:

$$(10.a) \quad c(I^d) = [(d + 1)/2]\binom{d}{[d/2]}.$$

It follows that $k(I^d) \geqslant a d^{1/2}$ for a suitable constant $a > 0$ but the exact values of $k(I^d)$ are not known for $d \geqslant 4$. While it is easily seen that $k(I^d) \leqslant d$, a remarkable example of Paterson shows that $k(I^6) \leqslant 5$.

D. W. Barnette [4] has established the following remarkable theorem, in which $]x[$ denotes the least integer $\geqslant x$:

(10.b) For every d-polytope P $k(P) \geqslant]\log_2(d + 1)[$, with equality if P is the d-dimensional simplex.

Among open problems related to cuts and cut numbers we mention:

Is $c(P) = c(I^d)$ for every d-polytope P isomorphic to I^d?

What is $k(I^d)$, and what is the least value of $k(P)$ for P isomorphic to I^d?

Is $k(P) \geqslant 1 +]\log_2 d[$ for every centrally symmetric d-polytope P?

Additional results, problems and references may be found in Grünbaum [3].

Remarks added November 18, 1974.

Since the completion of the preceding survey, many new developments have occurred that would deserve to be included. In order to keep the length of the survey within reasonable bounds, we shall report here only on a few items that materially changed the situation reported above.

1. Brunel's conjecture (discussed on page 210) has been completely established; details and extensions are given in B. Grünbaum and J. Zaks, "The existence of certain planar maps," Discrete Math., 10(1974), 93–115.

2. A detailed account of Grinberg's condition for the non-existence of a Hamiltonian circuit (see above, page 211) appears in Chapter 7 of R. H. Honsberger, "Mathematical Gems" (Dolciani Math. Expositions No. 1, Math. Assoc. of America, 1973).

3. A far-reaching improvement of Kotzig's theorem was obtained in E. Jucovič, "Strengthening of a theorem about 3-polytopes," Geometriae Dedicata 3(1974), 233–237.

4. For results on acyclic colorings of planar graphs related to those discussed in (6.b) see J. Mitchem, "Every planar graph has an acylic 8-coloring," Duke Math. J. 41(1974), 177–181.

REFERENCES

A. Altshuler
  1. "Hamiltonian circuits in some maps on the torus," *Discrete Math.*,
     1 (1972), 299–314.

M. L. Balinski
  1. "On the graph structure of convex polyhedra in n-space," *Pacific J. Math.*, **11** (1961), 431–434.

D. W. Barnette
  1. "Trees in polyhedral graphs," *Canad. J. Math.*, **18** (1966), 731–736.
  2. "On p-vectors of 3-polytopes," *J. Combinat. Theory*, **7** (1969), 99–103.
  3. "A completely unambiguous 5-polyhedral graph," *J. Combinat. Theory*, **9** (1970), 44–53.
  4. "Cut numbers of convex polytopes," *Geometriae Dedicata*, **2** (1973), 49–50.

D. W. Barnette and B. Grünbaum
  1. "On Steinitz's theorem concerning convex 3-polytopes and on some properties of 3-connected graphs," *Lecture Notes in Math.*, **110**, Berlin: Springer-Verlag, 1969, 27–40.

G. Brunel
  1. "Sur quelques configurations polyédrales," *Procès-verbaux séances Soc. Sci. Phys. Nat. Bordeaux*, 1897–98, 20–23.

V. Chvátal
  1. "Tough graphs and Hamiltonian circuits," *Discrete Math.*, **5** (1973), 215–228.

R. A. Duke
  1. "Geometric embedding of complexes," *Amer. Math. Monthly*, **77** (1970), 597–603.
  2. "On the genus and connectivity of Hamiltonian graphs," *Discrete Math.*, **2** (1972), 199–206.

V. Eberhard
  1. *Zur Morphologie der Polyeder*, Leipzig: Teubner, 1891.

G. Ewald
  1. "Hamiltonian circuits in simplicial complexes," *Geometriae Dedicata*, **2** (1973), 115–125.

P. J. Federico
  1. "Enumeration of polyhedra: The number of 9-hedra," *J. Combinat. Theory*, **7** (1969), 155–161.
  2. "The number of polyhedra." (To appear.)

J. C. Fisher
  1. "An existence theorem for simple convex polyhedra," *Discrete Math.*, **7** (1974), 75–97.

R. Forcade
 1. "Smallest maximal matchings in the graph of the d-dimensional cube," *J. Combinat. Theory (B)*, **14** (1973), 153–156.
T. Gallai
 1. "Signierte Zellenzerlegungen. I," *Acta Math. Acad. Sci. Hungar.*, **22** (1971), 51–63.
E. J. Grinberg
 1. "Plane homogeneous graphs of degree three without Hamiltonian circuits," *Latvian Math. Yearbook* 4, Izdat. "Zinatne," Riga 1968, 51–58. [Russian].
B. Grünbaum
 1. *Convex polytopes*, London: Interscience, 1967.
 2. "Polytopes, graphs, and complexes," *Bull. Amer. Math. Soc.*, **76** (1970), 1131–1201.
 3. "How to cut all edges of a polytope?" *Amer. Math. Monthly*, **79** (1972), 890–895.
 4. "Acyclic colorings of planar graphs," *Israel J. Math.*, **14** (1973), 390–408.
 5. "Vertices missed by longest paths or circuits," *J. Combinat. Theory (A)*, **17** (1974), 31–38.
 6. "Matchings in polytopal graphs," *Networks*, 4 (1974), 175–190.
B. Grünbaum and H. Walther
 1. "Shortness exponents of families of graphs," *J. Combinat. Theory (A)*, **14** (1973), 364–385.
S. Jendrol' and E. Jucovič
 1. "Generalization of a theorem by V. Eberhard." (To appear.)
E. Jucovič
 1. "On the number of hexagons in a map," *J. Combinat. Theory (B)*, **10** (1971), 232–236.
T. P. Kirkman
 1. "On the representation of polyedra," *Philos. Trans. Roy. Soc. London*, **146** (1856), 413–418.
 2. "Question 6610," Math. Questions from the *Educational Times*, **35** (1881), 112–116.
V. Klee
 1. "A property of d-polyhedral graphs," *J. Math. Mech.*, **13** (1964), 1039–1042.
 2. "Convex polytopes and linear programming," *Proc. IBM Sci.*

Sympos. Combinat. Problems (Yorktown Heights, N. Y. 1964). White Plains, N. Y.: IBM Data Process. Divis., 1966, 123–158.

A. Kotzig
1. "Príspevok k teórii Eulerovských polyédrov," *Mat.-Fyz. Časopis Slovensk. Akad. Vied.*, **5** (1955), 101–113. [Slovak. Russian summary.]

D. G. Larman and C. A. Rogers
1. "Paths in the one-skeleton of a convex body," *Mathematika*, **17** (1970), 293–314.

L. A. Lyusternik
1. "Convex figures and polyhedra," *GITTL*, Moscow 1956. [Russian] English translations, New York: Dover, 1963; Boston: Heath, 1966.

J. Malkevitch
1. "Properties of planar graphs with uniform vertex and face structure," *Mem. Amer. Math. Soc.*, **99** (1970).
2. "3-valent 3-polytopes with faces having fewer than 7 edges," *Ann. New York Acad. Sci.*, **175** (1970), 285–286.

P. Mani
1. "Automorphismen von polyedrischen Graphen," *Math. Ann.*, **192** (1971), 279–303.

P. McMullen
1. "The maximum numbers of faces of a convex polytope," *Mathematika*, **17** (1970), 179–184.

A. I. Medyanik
1. "On certain combinatorial properties of primitive convex polyhedra," *Ukrain. Geometr. Sbornik 1972*, **12**, 94–105. [Russian]

J. W. Moon and L. Moser
1. "Simple paths on polyhedra," *Pacif. J. Math.*, **13** (1963), 629–631.

P. E. O'Neil
1. "Hyperplane cuts of an n-cube," *Discrete Math.*, **1** (1971), 193–195.

M. D. Plummer
1. "On the cyclic connectivity of planar graphs," in *Graph Theory and Applications*, Y. Alavi et al., eds. Berlin: Springer, 1972, 235–242.

D. A. Rowland
1. "An extension of Eberhard's Theorem," *M. Sci. Thesis*, Seattle: Univ. of Washington, 1968.

C. F. Sainte-Marie
  1. "Question 505," *Interméd. Math.* **2** (1895), 93; *ibid.* **8** (1901), 308.

V. Schlegel
  1. "Theorie der homogen zusammengesetzten Raumgebilde," *Nova Acta Leop.-Carol. Deutsch. Akad. Naturforscher.*, **44** (1883), 339–459.

E. Steinitz
  1. "Polyeder und Raumeinteilungen," *Enzykl. Math. Wiss.*, **3** (1922), *Geometrie*, 3AB12, 1–139.

E. Steinitz and H. Rademacher
  1. "Vorlesungen über die Theorie der Polyeder," Berlin: Springer, 1934.

P. G. Tait
  1. "Remarks on the colouring of maps," *Proc. Roy. Soc. Edinburgh*, **10** (1880), 501–503.

W. T. Tutte
  1. "On Hamiltonian circuits," *J. London Math. Soc.*, **21** (1946), 98–101.
  2. "A theorem on planar graphs," *Trans. Amer. Math. Soc.*, **82** (1956), 99–116.
  3. "A theory of 3-connected graphs," *Nederland. Akad. Wetensch. Proc. Ser. A 64 = Indagat. Math.*, **23** (1961), 441–455.
  4. "On the enumeration of planar maps," *Bull. Amer. Math. Soc.*, **74** (1968), 64–74.

P. Ungar
  1. "On diagrams representing maps," *J. London Math. Soc.*, **28** (1953), 336–342.

EIGENVALUES OF GRAPHS

*A. J. Hoffman**

1. INTRODUCTION

Let G be a graph. The adjacency matrix $A(G)$ of G is a matrix of 0's and 1's defined as follows: number the vertices of G from 1 to $n = |V(G)|$ in some order, and set

$$A = A(G) = (a_{ij}) = \left\{ \begin{array}{l} 1 \text{ if } i \text{ and } j \text{ are adjacent vertices} \\ 0 \text{ if } i \text{ and } j \text{ are not adjacent vertices} \end{array} \right\}.$$

Apart from the numbering of the vertices, the adjacency matrix contains in principle all information about the graph. It is therefore natural to ask if any of the tools and concepts from matrix theory can be useful in studying graphs.

Another widely studied "adjacency matrix" for a graph has 0 on the diagonal, and the off diagonal entries -1 or $+1$ depending on

*This work was supported (in part) by the Army Research Office under contract number DAHCO4-72-C-0023.

225

whether the corresponding vertices are adjacent or non-adjacent. A survey of properties and many interesting applications of these matrices is given in [35]. In this chapter, we will confine ourselves to adjacency matrices of the type described in the preceding paragraph.

In this article, we shall study some relations between properties of a graph and properties of the eigenvalues of its adjacency matrix. If A is any matrix of order n, the eigenvalues of A are the n roots $\lambda_1, \ldots, \lambda_n$ of the polynomial $\det(\lambda I - A)$, where I is the identity matrix. We shall confine ourselves to topics arising only in graph theory, but related questions in permutation groups, design of experiments, elliptic geometry, statistical mechanics, and other disciplines, have also arisen. We shall assume a bare minimum of knowledge of matrix theory, citing facts from that subject as needed.

The author has spent a number of years investigating the interplay between a graph and the spectrum (set of eigenvalues) of its adjacency matrix, but still is reluctant to assert strongly that the spectrum of the adjacency matrix offers much or little insight into the graph. Given an arbitrary question about a graph, it is probably true that knowledge of the spectrum tells you little. But for some questions, the spectrum tells a great deal.

2. TWO EASY QUESTIONS

In this section, we shall show that the questions of whether the graph is regular (i.e., every vertex has the same valence), and whether the graph is bipartite, are completely determined by the spectrum. These are easy to show, and will establish some of the tools for studying more complicated questions.

We assume the graph G has at least one edge, now and throughout the article. Since all diagonal entries of $A(G)$ are 0, we know that the sum of the eigenvalues of A is 0, since (for any matrix A), trace $A \equiv \sum_i a_{ii} = \sum_i \lambda_i$. Next, $A = A(G)$ is a real symmetric matrix, since $a_{ij} = 1$ if and only if $a_{ji} = 1$. Consequently, every eigenvalue of A is real, and we will use the notation $\lambda_1 \geqslant \lambda_2 \geqslant \cdots$ to denote the eigenvalues in descending order, and $\lambda^1 \leqslant \lambda^2 \leqslant \cdots$ to denote the eigenvalues in ascending order.

From the theory of real symmetric matrices, we use the following:

(2.1) *If B is any principal submatrix of A, $\lambda_i(B) \leqslant \lambda_i(A)$ and $\lambda^i(B) \geqslant \lambda^i(A)$ for all i at most the order of B.*

We shall only be concerned with real vectors, and use the notation $(x, y) = \Sigma x_i y_i$.

(2.2) *For any $x \neq 0$, $(Ax, x)/(x, x) \leqslant \lambda_1$. If equality occurs, $Ax = \lambda_1 x$. For any $x \neq 0$, $(Ax, x)/(x, x) \geqslant \lambda^1$. If equality occurs, $Ax = \lambda^1 x$.* (Rayleigh's)

Now we can prove [5].

(2.3) *For any G, with $|V(G)| = n$, $(1/n)\Sigma\lambda_i^2 \leqslant \lambda_1$. Equality occurs if and only if G is regular of valence λ_1.*

Proof: Since $A = A(G)$ is symmetric, trace $A^2 = \Sigma\lambda_i^2 = \Sigma d_i$, where d_i is the valence of vertex i. To see this, realize that row i of A contains exactly d_i 1's. Hence, $(A^2)_{ii} = d_i$.

Let $u = (1, \ldots, 1)$. Then $(Au, u)/(u, u) = \Sigma d_i/n$. Thus, (2.3) follows from the first sentence in (2.2).

Notice that, besides being real and symmetric, $A(G)$ has all entries nonnegative. From the theory of nonnegative matrices, we inherit. *(Perron-Frobenius)*

(2.4) $\lambda_1 \geqslant |\lambda_i|$ *for all i. If G is connected, $x \neq 0$, at least one coordinate of x positive and $Ax = \lambda_1 x$, then $x > 0$. Also, when G is connected, $\lambda_1 > \lambda_i$ for all $i > 1$.*

With this we can prove [15].

(2.5) *If G is bipartite, $\lambda_i = -\lambda^i$ for all i. If G is connected, and $\lambda_1 = -\lambda^1$, then G is bipartite.*

Proof: Assume G bipartite, so

$$A(G) = \begin{pmatrix} 0 & B \\ B^\tau & 0 \end{pmatrix}.$$

Let λ be an eigenvalue of A, x the corresponding eigenvector (i.e., $Ax = \lambda x$). Write $x = (y; z)$ so that $Ax = \lambda x$ is equivalent to $Bz = \lambda y$, $B^\tau y = \lambda z$. Then clearly $A(y; -z) = -\lambda(y; -z)$.

Next, assume G connected, $Ax = \lambda_1 x$, $x > 0$ (2.4). Let X be the diagonal matrix whose ith coordinate is x_i. Then $C = X^{-1}AX$ is a nonnegative matrix whose row sums are λ_1. Further, we know from matrix theory that C and A have the same spectrum, so $\lambda^1 = -\lambda_1$ is the least eigenvalue of C. Let $Cy = \lambda^1 y$, $y \neq 0$, and let i^* be defined by $|y_{i^*}| = \max_i |y_i|$. Then $\lambda^1 y_{i^*} = \sum_j c_{i^* j} y_j$. By taking absolute values of both sides, we obtain

$$\lambda_1 |y_{i^*}| = |\lambda^1 y_{i^*}| = |\sum_j c_{i^* j} y_j| \leqslant \sum_j c_{i^* j} |y_j| \leqslant \sum_j c_{i^* j} |y_{i^*}| = \lambda_1 |y_{i^*}|.$$

Since the term on the left equals the term on the right, it follows that for all vertices j adjacent to i^*, the corresponding y_j are the same, and have the same absolute value as $|y_{i^*}|$. From $\lambda^1 y_{i^*} = \sum_j c_{ij} y_j$, we conclude that these y_j have sign opposite to the sign of y_{i^*}. But since G is connected, this proves G is bipartite.

3.   REGULAR CONNECTED GRAPHS AND THEIR POLYNOMIALS

The most common application of the regularity of a graph in studies involving eigenvalues is through the polynomial of a graph [15]. Let G be a regular connected graph of valence d with distinct eigenvalues (we shall speak of eigenvalues of $A(G)$ as eigenvalues of G): $\alpha_1 = d > \alpha_2 > \alpha_3 > \cdots > \alpha_k$. If $|V(G)| = n$, let

$$P(x) = \frac{n \prod\limits_{2}^{k} (x - \alpha_i)}{\prod\limits_{2}^{k} (d - \alpha_i)}. \tag{3.1}$$

Then, if $A = A(G)$, $P(A) = J$, where J is the matrix all of whose entries are 1.

The proof relies on the fact that, if two real symmetric matrices commute, then they can be simultaneously diagonalized by an orthogonal transformation. This means that, if $AB = BA$, there is a matrix U such that $U^\tau = U^{-1}$, and $U^\tau AU$ and $U^\tau BU$ are both diagonal, respectively A_0 and B_0. The entries in A_0 are the eigenvalues of A and the entries in B_0 are the eigenvalues of B. Indeed, there is a pairing of the eigenvectors of A and B so that, for each i, the ith diagonal entry of A_0 and the ith diagonal entry of B_0 share a common eigenvector.

Now A regular of valence d shows that $AJ = JA$, where $A = A(G)$. By (2.3) and (2.4), the distinct eigenvalues of A are d, $\alpha_2, \ldots, \alpha_k$, and d has multiplicity one. The distinct eigenvalues of J are n, and 0, and n has multiplicity one and is paired with d, with the common eigenvector $u = (1, \ldots, 1)$ of J and A. Let $UA_0U^\tau = A$, $UJ_0U^\tau = J$. Since $P(A_0) = J_0$, it follows that

$$UP(A_0)U^\tau = UJ_0U^\tau,$$

but $UP(A_0)U^\tau = P(UA_0U^\tau) = P(A)$, and $UJ_0U^\tau = J$.

It is easy to show that, if $P(A) = J$ for any polynomial, where $A = A(G)$, then G is regular and connected, and that (3.1) is the polynomial of least degree satisfying $P(A) = J$.

An interesting application of the polynomial of a graph occurs in the study of Moore graphs (d, k) [14]. A Moore graph of diameter k and valence d is a regular connected graph of valence d whose girth (the smallest cycle of length at least three) is $2k + 1$. Let us study what Moore graphs can exist apart from $2k + 1 -$ gons $(d = 2)$.

We consider the case $k = 2$. The definitions tell us that $|V(G)| = d^2 + 1$, and $A = A(G)$ satisfies

$$A^2 + A - (d - 1)I = J,$$

so from (3.1), the eigenvalues of A are d (with multiplicity 1) and numbers α that satisfy $\alpha^2 + \alpha - (d - 1) = 0$, so

$$\alpha_1 = \frac{-1 + \sqrt{4d - 3}}{2}, \quad \alpha_2 = \frac{-1 - \sqrt{4d - 3}}{2},$$

with respective multiplicities m_1 and m_2. Suppose $4d - 3$ is not a square. Then α_1 and α_2 are algebraic conjugates and must have the same multiplicity, so $m_1 = m_2 = d^2/2$. Since trace $A = 0$, we have $0 = d - d^2/2$, so $d = 2$. If $4d - 3 = s^2$, then m_1 and m_2 satisfy $m_1 + m_2 = d^2$, and trace

$$A = 0 = d + \frac{-1+s}{2} m_1 + \frac{-1-s}{2} m_2.$$

Expressing everything in terms of s, we find that s is a root of a monic polynomial whose constant term is 15. Therefore, $s = 1, 3, 5, 15$. The case $s = 1$ is impossible. The case $s = 3$ corresponds to the Petersen graph (figure 1).

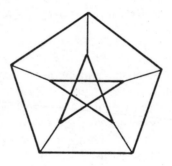

FIG. 1

The case $s = 5$ corresponds to a graph of 50 vertices completely described in [14]. See also [2]. Whether there is a graph corresponding to the case $s = 15$ is unknown.

To study the problem for more general k, let P_k be the polynomial of the graph. If we set $P_0 = 1$, $P_1 = 1 + x$, then

$$P_k = xP_{k-1} - (d - 1)P_{k-2}. \tag{3.2}$$

Bannai and Ito [1] and also Damerell [7] have shown from a detailed examination of (3.2) and properties of the spectrum of the graph that no Moore graphs exist for $k > 2$.

4. LINE GRAPHS

If G is a graph, the line graph of G (denoted by $L(G)$) is a graph whose vertices are the edges of G, with two vertices of $L(G)$ adjacent if the corresponding edges of G have one common vertex. The spectrum of $A(L(G))$ has the following remarkable property:

$$\lambda^1(L(G)) \geqslant -2. \tag{4.1}$$

To see this, we use

If K is any real (rectangular) matrix, $\lambda^1(KK^\tau) \geqslant 0$. (4.2)

To prove (4.1), let K be the (0, 1) matrix whose rows correspond to the edges of G, columns to vertices of G, with the entry in K equal to 1 if and only if the edge contains the vertex. Then $\lambda^1(KK^\tau) \geqslant 0$ by (4.2). But $KK^\tau = 2I + A(L(G))$, whence (4.1) follows.

Inspired either by (4.1) or by problems arising naturally in other ways, there have been many studies of the line graphs and their spectra. For example, $L(K_n)$ is characterized by its spectrum if $n \neq 8$ (there are three exceptions if $n = 8$, [3], [4], [12], [13]); $L(K_{n,n})$ is characterized by its spectrum if $n \neq 4$ (there is one exception if $n = 4$, [32]); $L(SBIBD(v, k, \lambda))$ is characterized by its spectrum unless $v = 4$, $k = 3$, $\lambda = 2$ (one exception, [18]); etc. To give the flavor of this work, let us look at $L(K_{n,n})$ from our viewpoint (which was not used originally). First, we determine the distinct eigenvalues of its spectrum.

If K is the edge-vertex incidence matrix of $K_{n,n}$, we have

$$KK^\tau = 2I + A(L(K_{n,n})), \qquad K^\tau K = nI + A(K_{n,n}).$$

By a well-known theorem of matrix theory, KK^τ and $K^\tau K$ have the same non-zero eigenvalues, and 0 is also an eigenvalue of both. Now the distinct eigenvalues of $A(K_{n,n})$ are $n, 0, -n$, so the distinct eigenvalues of $K^\tau K$ are $2n, n, 0$; so the distinct eigenvalues of $L(K_{n,n})$ are

$$2n - 2, n - 2, -2. \tag{4.3}$$

Assume G a regular connected graph on n^2 vertices with (4.3) as distinct eigenvalues. We wish to prove that, if $n \neq 4$, G is $L(K_{n,n})$.

To do this, we first define $H \subset G$ to mean that the graph H is obtained from G by choosing a subset of its vertices, and all edges of G which join vertices in the subset. Another way of expressing the same idea is that $A(H)$ is a principal submatrix of $A(G)$. Thus from (2.1), we have

$$\text{if } H \subset G, \quad \lambda^1(H) \geq \lambda^1(G). \tag{4.4}$$

In addition,

If G regular, $H \subset G$, $\lambda^1(H) = \lambda^1(G)$, $A(H)x = \lambda^1(H)x$, then $\Sigma x_i = 0$; also, if $v \in V(G) - V(H)$, then $\Sigma' x_i = 0$, where Σ' is over all vertices in H adjacent to v. (4.5)

Proof of (4.5): Let $y = (x; 0 \ldots 0)$ be the vector of $|V(G)|$ coordinates obtained by setting $y_i = x_i$ for $i \in V(H)$, $y_i = 0$ for $i \in V(G) - V(H)$. Then $(A(G)y, y)/(y, y) = (A(H)x, x)/(x, x) = \lambda^1(x, x)/(x, x) = \lambda^1$. Therefore, by (2.2), y is an eigenvector of $A(G)$ corresponding to the eigenvalue λ^1. But $u = (1, \ldots, 1)$ is an eigenvector of $A(G)$ corresponding to the eigenvalue d, where d is the valence of each vertex of G. A theorem about real symmetric matrices says that eigenvectors corresponding to different eigenvalues are orthogonal (their inner product is 0). Thus $(y, u) = (x, u) = \Sigma x_i = 0$. Further, the fact that y is an eigenvector of $A(G)$ says $\Sigma_j a_{vj} y_j = 0 = \Sigma' x_i$.

From (4.3), we know $-2 = \lambda^1(G)$. Then the following graphs cannot be subgraphs of G (we have attached coordinates of the eigenvector x corresponding to the eigenvalue -2 to the corresponding vertices).

(4.6)   (4.7)   (4.8)   (4.9)

Graphs (4.6) and (4.7) are excluded because the corresponding eigenvectors have the sum of coordinates not 0, violating the first part of (4.5). Graphs (4.8) and (4.9) violate the second part of (4.5). Now we show that (4.3) implies

$K_{1,3} =$

cannot be a subgraph of G if $n \neq 4$. This is as far as we go here. (Note $K_{1,3} \not\subset G$ is a necessary condition for G to be a line graph.) From then on, the proof that (4.3) characterizes $L(K_{n,n})$ if $n \neq 4$ need not exploit eigenvalues.)

From (4.3) and (3.1), we conclude that $P(A) = J$, where

$$P(x) = n^2(x - n + 2)(x + 2)/2n^2$$

$$= \tfrac{1}{2}(x^2 + (4 - n)x - 2(n - 2)).$$

This can be rewritten as

$$A^2 = 2(n - 1)I + (n - 2)A + 2(J - I - A). \qquad (4.10)$$

Assume $K_{1,3} \subset G$, with the vertices of $K_{1,3}$ labelled 0, 1, 2, 3, so that 0 is adjacent to each of $\{1, 2, 3\}$, no two of which are adjacent to each other. Since $(4.6) \not\subset G$, every vertex of G adjacent to 0 other than 1, 2, 3 must be adjacent to at least one of $\{1, 2, 3\}$; it cannot be adjacent to all of $\{1, 2, 3\}$, for that would make $(4.7) \subset G$. So the $2(n - 1) - 3$ vertices adjacent to 0 other than 1, 2, 3 can be thought of as belonging to six sets S_1, S_2, S_3, S_{12}, S_{13}, S_{23}, where S_i consists of vertices adjacent to i but not to j or k, S_{ij} consists of vertices adjacent to i and j but not to k. From (4.10), if two different vertices of G are not adjacent, there are exactly two vertices adjacent to both. Since $(4.8) \not\subset G$ and $(4.9) \not\subset G$, it follows that $|S_{ij}| = 1$. Thus

$$2(n - 1) - 3 = \sum |S_i| + \sum |S_{ij}|$$

$$= \sum |S_i| + 3, \quad \text{or} \quad \sum |S_i| = 2n - 8.$$

Also from (4.10), if two vertices of G are adjacent, there are exactly $(n - 2)$ vertices adjacent to both. Thus

$$|S_i| + |S_{ij}| + |S_{ik}| = |S_i| + 2 = n - 2.$$

Therefore $\Sigma|S_i| = 3(n - 4)$. Therefore, $3n - 12 = 2n - 8$ or $n = 4$. Hence, $n \neq 4$ implies $K_{1,3} \not\subset G$.

One can show that if G is a regular connected graph of valence > 12, $\lambda^1(G) = -2$, then $K_{1,3} \not\subset G$. One can also show that, if G is a regular connected graph of valence > 16, $\lambda^1(G) = -2$, then G is a line graph or a cocktail party graph (the complement of the graph consisting of n disjoint edges). The numbers 12 and 16 are best possible. More generally, one can show there is an integer function f defined on the open interval $(-1 - \sqrt{2}, -1)$ such that if G is any connected graph with $\lambda^1(G)$ in the interval, and the minimum valence of the vertices of G exceeds $f(\lambda^1(G))$, then G is a "generalized" line graph (described in [22]). In addition, all regular connected graphs with three distinct eigenvalues and $\lambda^1 = -2$ are known [31]. Also, if $m + n$ is sufficiently large, $L(K_{m,n})$ is characterized by its spectrum, except for an infinite number of exceptions corresponding precisely to line graphs associated with certain symmetric Hadamard matrices [9]. There are other results in this spirit for line graphs associated with Steiner triple systems and other designs [8].

(**Added in galley**: P. J. Cameron, J. M. Goethals, J. J. Seidel and E. E. Shult "Line graphs, root systems, and elliptic geometry" (to appear), using the concept of root systems from Lie theory, have subsumed and improved many of the results of this section.)

5. SIGNIFICANCE OF BOUNDS ON THE SPECTRUM

The significance of $\lambda^1(G) \geqslant -2$ has been roughly summarized, but almost nothing is known about the significance of other specific lower bounds. What is known is the meaning of the existence of some lower bound [25]. Let us first explain the question.

Let **G** be any (infinite) family of graphs. The following statement about **G** may be true or false:

(5.1) *There exists a number* λ *such that, for all* $G \in \mathbf{G}$, $\lambda^1(G)$ $> \lambda$.

We shall describe "local" and "global" properties of the graphs $G \in \mathbf{G}$ equivalent to (5.1). First, local properties.

Let us define H_n to be the graph on $2n + 1$ vertices in which one vertex is not adjacent to n other vertices, otherwise all vertices are adjacent. Then $\lambda^1(H_n) \to -\infty$. To see this, first note

$$A(H_n) = \begin{pmatrix} 0 & u & 0 \\ u' & J - I & J \\ 0 & J & J - I \end{pmatrix}.$$

Clearly, -1 is an eigenvalue of multiplicity at least $2n - 2$, with corresponding eigenvectors of the form all coordinates 0 except two coordinates $+1$, -1. It follows that the eigenvectors for the three remaining eigenvalues have all coordinates the same on each block; i.e., are of the form $(0; x, \ldots, x; y, \ldots, y)$. Thus the three remaining eigenvalues are eigenvalues of the 3×3 matrix

$$M = \begin{pmatrix} 0 & n & 0 \\ 1 & n - 1 & n \\ 0 & n & n - 1 \end{pmatrix}.$$

Now $\det(xI - M) = x^3 - 2(n - 1)x^2 + (1 - 3n)x + n(n - 1)$ $\equiv M_n(x)$. By Descartes' rule of signs, there is exactly one negative root of $M(x)$, say α_n. Further, this $\alpha_n < -1$, since H_n contains $K_{1,2}$ and $\lambda^1(K_{1,2}) = -\sqrt{2} < -1$ and we apply (4.4). Thus $\alpha_n = \lambda^1(H_n)$. Since $H_n \subset H_{n+1}$, it follows from (4.4) that $\lambda^1(H_n)$ is a non-increasing sequence. If it is false that $\lambda^1(H_n) \to -\infty$, then $\lambda^1(H_n) \to \bar{\lambda} \neq -\infty$. Since $M_n(\lambda^1(H_n)) = 0$, it follows that $M_n(\lambda^1(H_n))/n(n - 1) = 0$. But $f_n(x) \equiv M_n(x)/n(n - 1)$ is a polynomial which is a sum of terms each of which approaches a limit as $\lambda^1(H_n) \to \bar{\lambda}$. Therefore, $\lim f_n(\lambda^1(H_n)) = 0 = \lim_{n \to \infty} f_n(\bar{\lambda})$ $= 1 \neq 0$. This contradiction shows $\lambda^1(H_n) \to -\infty$.

Also, $\lambda^1(K_{1,n}) = -\sqrt{n} \to -\infty$. Thus, (4.4) shows that (5.1) implies

(5.2) *There exists an integer* n_0 *such that, for all* $G \in \mathbf{G}$, $H_{n_0} \not\subset G$, $K_{1,n_0} \not\subset G$.

It is a remarkable fact that (5.2) implies (5.1), but we will not prove it here. But (5.2) is the local property of graphs equivalent to (5.1) which we promised. Now for the global property.

We first define a distance between two graphs G and H where $V(G) = V(H)$. For each vertex $i \in V(G) = V(H)$, let e_i be the larger of the two numbers {number of edges in G on i which are not in H, number of edges in H on i which are not in G}. Then $d(G, H) = \max_i e_i$. Now consider the following statement about G:

(5.3) *There exists an integer L such that, for each $G \in G$, there is a graph H with $V(G) = V(H)$ such that $d(G, H) < L$, and H contains a distinguished family of cliques $K^1, K^2, \ldots$, satisfying*

(a) *Every edge of H is in at least one of the distinguished cliques.*
(b) *Every vertex of H is in at most L of the distinguished cliques.*
(c) *Any two distinguished cliques have at most L common vertices.*

We assert that (5.3) is also equivalent to (5.1). The full proof of the equivalence of (5.1) and (5.3) consists of showing (5.1) implies (5.2) (which we have done), (5.2) implies (5.3) (a lengthy argument, entirely graph-theoretical, which we omit), and (5.3) implies (5.1), which we now indicate. But first some facts about matrices:

(5.4) *If A, B, and C are real symmetric matrices and $A - B = C$, then $\lambda^i(A) - \lambda^i(B) \leq \lambda_1(C)$.*

(5.5) *If A is any matrix and abs A is defined by $(\text{abs } A)_{ij} = |a_{ij}|$ for all i, j, then each eigenvalue of A is in absolute value at most the largest eigenvalue of abs A.*

(5.6) *If A is any nonnegative matrix, then its largest eigenvalue is at most the largest row sum.*

From $d(G, H) < L$, we conclude that there exist graphs $\tilde{G}$ and $\tilde{H}$ such that

$$A(G) + A(\tilde{G}) = A(H) + A(\tilde{H}),$$

where $\tilde{G}$ and $\tilde{H}$ have no common edges, and each has valence less than L. Using (5.4), we have

$$- \lambda^1(G) + \lambda^1(H) \leqslant \lambda_1\left(-A(\tilde{H}) + A(\tilde{G}) \right)$$

$$\leqslant \lambda_1\left(A(\tilde{H}) + A(\tilde{G}) \right) \text{ (by (5.5))} < 2L \text{ (by 5.6))}.$$

Therefore, all we need show is that $\lambda^1(H)$ is bounded from below by a function of L.

Let M be the incidence matrix of points versus distinguished cliques of H. Then $MM^T = A(H) + S$, where S is a nonnegative matrix (by (5.3) (a)). Every diagonal entry of S is at most L (by (5.3)(b)), and the sum of the off diagonal entries in each row is at most $(L - 1)\binom{L}{2}$ (using (5.3)(b) and (c)). Thus

$$0 \leqslant \lambda^1(MM^T) \leqslant \lambda^1(H) + \lambda_1(S)$$

$$\leqslant \lambda^1(H) + L + (L - 1)\binom{L}{2} \quad \text{(using (5.6))},$$

which was to be proven.

One can raise analogous questions for each eigenvalue: upper bounds on λ_i? lower bounds on λ^i? For λ_2, the corresponding local and global properties have been found in [26]. For λ_1, the local and global equivalences are easy. Nothing has yet been found about other λ_i and λ^i.

6. IMBEDDINGS OF GRAPHS

If $G \subset H$, we know $\lambda_i(G) \leqslant \lambda_i(H)$, $\lambda^i(G) \geqslant \lambda^i(H)$. Suppose G given, and we wish to find the smallest amount by which λ_i or λ^i changes when G is imbedded in H satisfying certain conditions. The most interesting results on this question so far deal with changes in λ^1 and λ_2 when every vertex of H is required to have large valence. The general problem can be stated:

Define $d(G)$ to be the minimum valence of the vertices of G, and

$$\mu_i(G) = \lim_{d \to \infty} \inf_{\substack{H \supset G \\ d(H) > d}} \lambda_i(H)$$

$$\mu^i(G) = \lim_{d \to \infty} \sup_{\substack{H \supset G \\ d(H) > d}} \lambda^i(H).$$

A fact about nonnegative matrices asserts that the largest eigenvalue is at least the minimum row sum. Hence, $d(H)$ large implies $\lambda_1(H)$ large, so $\mu_1 = \infty$. Every other μ_i and μ^i is finite, however, because μ_2 and μ^1 are finite. In fact, there are formulas for μ_2 and μ^1. Let $|V(G)| = m$. Let $\mathbf{C}^1$ be the set of rectangular $(0, 1)$ matrices C with m rows satisfying:

(6.1) every row sum is positive, and
(6.2) if any column C is deleted, (6.1) is false.

Let $\mathbf{C}_2$ be the set of rectangular $(0, 1)$ matrices C with m rows and at least two columns, satisfying (6.1), and, if C has more than two columns, (6.2).
Then, letting $A = A(G)$,

$$\mu^1 = \max_{C \in \mathbf{C}^1} \lambda^1(A - CC^\tau). \tag{6.3}$$

$$\mu_2 = \min_{C \in \mathbf{C}_2} \lambda_1(A - C(J - I)^{-1}C^\tau), \quad \text{if } |V(G)| \geq 2. \tag{6.4}$$

We shall prove here

$$\mu^1 \geq \max_{C \in \mathbf{C}^1} \lambda^1(A - CC^\tau) \qquad ([19]) \tag{6.3}$$

and merely indicate the proof of the reverse inequality.

Let G and $C \in \mathbf{C}^1$ be given. Construct a graph $H(n)$ as follows: Assume C has columns $C_1, \ldots, C_t$. The vertices of $H(n)$ consist of the vertices of G and the vertices of t cliques $K_n^1, \ldots, K_n^t$ of n vertices. The edges in G and the edges in each K_n^j are edges of $H(n)$. In addition, vertex i of G is adjacent to every vertex of K_n^j if

$c_{ij} = 1$. If $c_{ij} = 0$, vertex i of G is adjacent to no vertex of K_n^j. Finally, if $j \neq k$, each vertex of K_n^j is not adjacent to each vertex of K_n^k. Thus,

$$A(H(n)) = \begin{pmatrix} A(G) & D \\ D^\tau & E \end{pmatrix},$$

where D is a matrix of m rows and nt columns, obtained by substituting $u = (1, \ldots, 1)$ for each 1 in C, and E is the direct sum of t matrices of order n, each of which is $J - I$.

We first assume $\lambda^1(G) < -1$. Now -1 is an eigenvalue of $A(H(n))$ of multiplicity at least $t(n - 1)$. Using the same arguments that arose in consideration of H_n in §5, the remaining eigenvalues of $H(n)$ are the eigenvalues of the matrix of order $m + t$

$$B(n) = \begin{pmatrix} A(G) & nC \\ C^\tau & (n-1)I \end{pmatrix}.$$

Since $\lambda^1(A) < -1$, $\lambda^1(B(n)) = \lambda^1(H(n))$. Since $H(n) \subset H(n+1)$, $\lambda^1(H(n))$ is nonincreasing. Further, let $H'(n)$ be the graph obtained from $H(n)$ by first deleting all edges in G, then putting in all edges of G joining all vertices i, k such that $c_{ij} = c_{kj} = 1$ for some $j = 1, \ldots, t$. Then $H'(n)$ satisfies (5.3) with $L = (|V(G)|)$. Thus $\lambda^1(H(n))$ is bounded from below and must approach a limit.

Now $\lambda^1(H(n))$ is the least root of $\det(\lambda I - B(n))$. A theorem of analysis (Hurwitz's theorem) tells us that if $P(n, x)$ is a polynomial in x, each of whose coefficients is a polynomial in n, then the limit points (as $n \to p$), if any exists, of the roots of $P(n, x)$ (for fixed n) are precisely the roots of $Q(x)$, where $Q(x)$ is the coefficient of the highest power of n when $P(n, x)$ is expressed as a polynomial in n. If we apply Laplace's expansion to $\det(\lambda I - B(n))$, it is clear that $Q(\lambda)$ is

$$\det\begin{pmatrix} \lambda I - A & -C \\ -C^\tau & -I \end{pmatrix}.$$

This determinant is 0 if and only if there exists a vector $(x; y)$ not

0, such that

$$(\lambda I - A)x - Cy = 0$$

$$- C^\tau x - y = 0$$

or

$$(\lambda I - (A - CC^\tau))x = 0,$$

which means λ is an eigenvalue of $A - CC^\tau$. Clearly, $\lim \lambda^1(H(n))$ = $\lambda^1(A - CC^\tau)$, which was to be proven.

There remains the case when $\lambda^1(G) \geq -1$. Let us first observe that for a graph H with at least one edge $\lambda^1(H) \leq 1$, since

$\lambda^1 \begin{pmatrix} 0 & 1 \\ 1 & 0 \end{pmatrix} = -1$, and we apply (4.4). Therefore $\mu^1(G) \leq -1$. Observe also that $\lambda^1(K) = -1$ for K a clique. It follows that $\mu^1(G)$ = -1 for G a union of disjoint cliques, since each clique could be expanded to arbitrarily large size. And this can be realized by letting each column of C correspond to the cliques ($=$ components). And if there is at least one component of G not a clique, then $K_{1,2} \subset G$, so $\lambda^1(G) \leq \lambda^1(K_{1,2}) = -\sqrt{2} < -1$.

To prove the reverse of (6.5) is more delicate. One shows using Ramsey's theorem and an analogous theorem for bipartite graphs that, if $H \supset G$ is a graph in which every vertex has large valence and $\lambda^1(H)$ is bounded from below, then for some $C \in \mathbf{C}^1$, there is a large n such that $G \subset H(n) \subset H$, whence (4.4) applies.

7. PARTITIONING OF GRAPHS

We will consider here only partitionings of the set of vertices ([20], [21], [6]), (some results on edge partitionings are given in [23]). The most famous partitioning of the vertices occurs in the coloring of a graph. A coloring of a graph is an assignment of colors to the vertices so that adjacent vertices have different colors. The symbol $\chi(G)$ stands for the smallest number of colors required. We shall easily derive upper and lower bounds for $\chi(G)$

from its spectrum

$$1 + \frac{\lambda_1(G)}{\overline{\lambda}^1(G)} \leqslant \chi(G) \leqslant 1 + \lambda_1(G). \qquad (7.1)$$

To prove the left inequality of (7.1), we use the following (not so well-known) theorem of matrix theory:

(7.2) *If*

$$A = \begin{pmatrix} A_{11}A_{12} & \cdots & A_{1t} \\ \vdots & & \\ A_{t1} & \cdots & A_{tt} \end{pmatrix}$$

is a partitioning of the real symmetric matrix A into blocks, with each diagonal block square, then

$$\lambda_1(A) + \sum_{i=1}^{t-1} \lambda^i(A) \leqslant \sum_{i=1}^{t} \lambda_1(A_{ii}).$$

Now if $A = A(G)$, $\chi(G) = t$, we can arrange so each $A_{ii} = 0$, whence $\lambda_1(A_{ii}) = 0$. This yields the left side of (7.1).

To prove the right side, discard vertices of G to obtain a graph H such that $\chi(H) = \chi(G)$, but discarding any other vertex lowers the coloring number. Then each vertex of H must be adjacent to at least $\chi - 1$ other vertices. Therefore, every row sum of $A(H)$ is at least $\chi - 1$. But $\lambda_1(G) \geqslant \lambda_1(H) \geqslant \chi(G) - 1$, which yields the right side of (7.1). [34]

One can also derive an upper bound for $\chi(G)$ in terms of the number of eigenvalues of G which are at most -1 (see [23]), and also prove

$$\chi(G) \geqslant \frac{|V(G)|}{|V(G)| - \lambda_1} \quad [6],$$

but the trouble with (7.1) and other spectral bounds is that they cannot be sharp. The lower bound is sharp for $\chi = 2$ (see § 2) and for line graphs and for cliques. The upper bound is sharp for cliques and many other cases. But given any $N > 3$, one can find two graphs G and H with identical spectra, $\chi(G) = 3$, $\chi(H) > N$ (see [23]).

It is interesting to look at the following vertex partition. Call vertices a and b equivalent if $a = b$ or if $a \neq b$ and every other vertex is adjacent to both or adjacent to neither. Let $e(G)$ be the number of equivalence classes. Now the number of equivalence classes is *not* determined by the spectrum. For example,

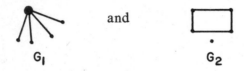

each has spectrum $(2, 0, 0, 0, -2)$, $e(G_1) = 2$, $e(G_2) = 3$. But a very rough order of magnitude of $e(G)$ is determined by the spectrum in the following sense:

For $a \leq b$, define $\Lambda(a, b)(G)$ to be the number of eigenvalues of G each of which is at most a or at least b. Then there exists a function f such that, for all graphs,

$$\Lambda(a, b)(G) \leq e(G) \leq f(\Lambda(a, b)(G)), \qquad (7.2)$$

where $a = -2$, $b = (\sqrt{5} - 1)/2$, or $a = (-\sqrt{5} - 1)/2$, $b = 1$. Further, in (7.2) we cannot replace a by $a - \epsilon$ or b by $b + \epsilon$ and have (7.2) remain true.

To prove the left inequality in (7.2), we first observe that each equivalence class consists of a clique or an independent set. A clique of size n_i produces $n_i - 1$ eigenvalues -1. An independent set of size n_i produces $n_i - 1$ eigenvalues 0. Thus $e(G) \geq$ number of eigenvalues of G which are not 0 or $-1 \geq \Lambda(a, b)(G)$ for the (a, b) pairs cited in (7.2).

To prove the right-hand ride, we use Ramsey's theorem to prove that, if $e(G)$ is large, then $A(G)$ contains a subgraph H with

$$A(H) = \begin{pmatrix} M & B \\ B^T & N \end{pmatrix},$$

where ord M = ord N = large, $M = 0$ or $J - I$, $N = 0$ or $J - I$, $B = I$ or $J - I$ or a triangular matrix T. We omit the subsequent calculations, which are based on knowledge of the eigenvalues of M, N, and $\begin{pmatrix} 0 & B \\ B^T & 0 \end{pmatrix}$ and applications of (5.4).

8. SOME QUESTIONS

What numbers can be eigenvalues of graphs? It is clear that any eigenvalue of a graph must be a totally real algebraic integer (a real algebraic integer all of whose conjugates are real). One can also show that every eigenvalue of a symmetric matrix of rational integers is an eigenvalue of a graph. So the question can be rephrased: is every totally real algebraic integer an eigenvalue of a symmetric matrix of rational integers? It is known that every totally real algebraic number is an eigenvalue of a symmetric matrix of rational numbers.

What numbers can be limit points of eigenvalues of graphs? By this we mean: fix i, and let G_1, G_2, ..., be a sequence of graphs such that $\lambda_i(G_j)$ (or $\lambda^i(G_j)$) approaches a limit $\bar{\lambda}_i$ (or $\bar{\lambda}^i$) as $j \to \infty$. We know only the following sketchy information: all limit points of the set of λ_1's which are at most $\sqrt{2 + \sqrt{5}}$ [24] and all limit points of the set of λ^1's which are at least -2.

Another question is: how do the eigenvalues change if, starting with a given graph G, we consider all graphs homeomorphic to G? All that is known are formulas for the inf and sup of $\lambda_1(H)$, as H ranges over all graphs homeomorphic to G.

BIBLIOGRAPHY

We confine ourselves to references dealing explicitly with the subjects treated here.

1. Bannai, E., and T. Ito, "On finite Moore graphs," *J. Fac. Sci.*, Univ. of Tokyo, Sect. IA, **20** (1973), 191–208.

2. Benson, C. T., and N. E. Losey, "On a graph of Hoffman and Singleton," *J. Combinatorial Theory*, **11B** (1971), 67–79.

3. Chang, L. C., "The uniqueness and non-uniqueness of the triangular association scheme," *Sci. Record*, **3** (1959), 604–613.

4. ——, "Association schemes of partially balanced block designs with parameters $v = 28$, $n_1 = 12$, $n_2 = 15$ and $p_{11}^2 = 4$," *ibid.* **4** (1960), 12–18.

5. Collatz, L. and U. Sinogowitz, "Spektren endlicher Grafen," *Abh. Math. Sem. Univ. Hamburg*, **21** (1957), 63–77.

6. Cvetkovic, D. M., "Chromatic number and the spectrum of a graph," *Pub. Inst. Math. (Beograd)*, **14**, 28 (1972), 25–38.

7. Damerell, R., "On Moore graphs," *Proc. Cambridge Philos. Soc.*, **74** (1973), 227–236.

8. Doob, M., "Graphs with a small number of distinct eigenvalues," *Ann. New York Acad. Sci.*, **175**, 1, (1970), 104–110.

9. ——, "On characterizing certain graphs with four eigenvalues by their spectra," *Linear Algebra and Appl.*, **3** (1970), 461–482.

10. Finck, H. J., and G. Grohmann, "Vollständiges Produkt, chromatische Zahl und charakteristisches Polynom regulärer Graphen I," *Wiss. Z. Techn. Hochsch. Ilmenau*, **11** (1965), 1–3.

11. ——, "Vollständiges Produkt, chromatische Zahl und characteristisches Polynom regulärer Graphen II," *ibid.* **11** (1965), 81–87.

12. Hoffman, A. J., "On the uniqueness of the triangular association scheme," *Ann. Math. Statist.*, **31** (1960), 492–497.

13. ——, "On the exceptional case in a characterization of the arcs of a complete graph," *IBM J. Res. Develop.*, **4** (1960), 487–496.

14. Hoffman, A. J., and R. R. Singleton, "On Moore graphs with diameters 2 and 3," *ibid.* (1960), 497–504.

15. Hoffman, A. J., "On the polynomial of a graph," *Amer. Math. Monthly*, **70** (1963), 30–36.

16. Hoffman, A. J., and D. K. Ray-Chaudhuri, "On the line graph of a finite plane," *Canad. J. Math.*, **17** (1965), 687–694.

17. Hoffman, A. J., "On the line graph of a projective plane," *Proc. Amer. Math. Soc.*, **16** (1965), 297–302.

18. Hoffman, A. J. and D. K. Ray-Chaudhuri, "On the line graph of a symmetric balanced incomplete block design," *Trans. Amer. Math. Soc.*, **116**, 4 (1965), 238–252.

19. Hoffman, A. J., "The change in the least eigenvalue of the adjacency

matrix of a graph under imbedding," *SIAM J. Appl. Math.*, **17** (1969), 664–671.

20. ——, "On eigenvalues and colorings of graphs" in *Graph Theory and Its Applications*, New York: Academic Press, 1970, 79–91.

21. Hoffman, A. J., and L. Howes, "On eigenvalues and colorings of graphs II," *Ann. New York Acad. Sci.*, **175** (1970), 238–242.

22. Hoffman, A. J., "$-1 - \sqrt{2}$?," in *Combinatorial Structures and Their Applications*, New York: Gordon and Breach, 1970, 173–176.

23. ——, "Eigenvalues and edge partitionings of graphs," *Linear Algebra and Appl.*, **5** (1972), 137–146.

24. ——, "On limit points of spectral radii of nonnegative symmetric integral matrices" in *Graph Theory and Its Applications*, New York: Springer Verlag, 1972, 165–172.

25. ——, "On spectrally bounded graphs," in *A Survey of Combinatorial Theory*, Amsterdam: North-Holland, 1973, 277–284.

26. Howes, L., "On subdominantly bounded graphs—summary of results," in *Recent Trends in Graph Theory, Lecture Notes in Mathematics* **186** New York: Springer Verlag, 1971, 181–183.

27. Kraus, L. L., and D. M. Cvetkovic, "Evaluation of a lower bound for the chromatic number of the complete product of graphs," *Univ. Beograd Publ. Elektrotehn. Fak. Ser. Mat. Fiz.*, **357–380**, 63–68.

28. Ray-Chaudhuri, D. K., "Characterization of line graphs," *J. Combinatorial Theory*, **3** (1967), 201–214.

29. Sachs, H., "Über Teiler, Faktoren und characteristische Polynome von Graphen, Teil I," *Wiss. Z. Techn. Hochsch. Ilmenau*, **12** (1966), 7–12.

30. ——, *ibid*, Teil II, **13** (1967), 405–412.

31. Seidel, J. J., "Strongly regular graphs with $(-1, 1, 0)$-adjacency matrix having eigenvalue 3," *Linear Algebra and Appl.*, **1** (1968), 281–298.

32. Shrikhande, S. S., "The uniqueness of the L_2 association scheme," *Ann. Math. Statist.*, **30** (1959), 781–798.

33. Singleton, R., "On minimal graphs of maximum even girth," *J. Combinatorial Theory*, **1** (1966), 306–332.

34. Wilf, H. S., "The eigenvalues of a graph and its chromatic number," *J. London Math. Soc.*, **42** (1967), 330–332.

35. Seidel, J. J., "A survey of two-graphs," *Proc. Int. Coll. Teorie Combinatorie, Acc. Naz. Lincei*, Rome (1973).

ON THE AXIOMATIC FOUNDATIONS OF THE THEORIES OF DIRECTED LINEAR GRAPHS, ELECTRICAL NETWORKS AND NETWORK-PROGRAMMING*

George J. Minty

1. INTRODUCTION

Matroid-theory, founded by Hassler Whitney [41] in 1935, is an abstract combinatorial theory with ramifications into algebra (theory of linear dependence [41], lattice theory [36]), projective geometry [39], electrical network theory [18], switching theory [23], and linear programming. In this paper, it is proposed to develop

*The writing of this paper was partially supported by N. S. F. Grants 23830 and GP-3465. An abstract (under the same title [30]) was given at the Seventh Midwest Symposium on Circuit Theory, Ann Arbor, Michigan, May 4–5, 1964. The paper was presented substantially in its present form at the Symposium on Matroid Theory sponsored by the Applied Mathematics Division of the National Bureau of Standards at Washington, D. C., September 1–11, 1964.

matroid-theory from the beginning in such a way as to put the phenomenon of *duality* into the forefront by starting with a self-dual axiom-system, following the model of projective geometry. The paper is, however, also intended as an expository paper on matroid-theory for the electrical engineer, the graph-theorist, and in fact the general mathematician. It is complete for each of these audiences, and the reader may skip over any remark obviously intended for another class of readers.

One of the main contributions of this theory to electrical theory is the clarification of the "duality principle," which every electrical engineer knows, but few can state so exactly that it is *provable*. (Since this paper was written, the writer has become aware of the work of members of the Research Association for Applied Geometry (R.A.A.G.) on duality-theory. A representative of that group, M. Iri, has compared their point of view with the present one in the paper [18].) By the end of this paper, it will be so clear that we shall not even find it necessary to state it! Another purpose will be to help clarify the meaning of the contentions of G. Kron [21] that "graphs are illegitimate tools for studying network-theory" and that "1-networks have no nodes." The axiom-system we shall work with is highly appropriate for the treatment of matroids as generalizations of linear graphs; the axioms we shall work with (for a "graphoid") are rather easily seen to be satisfied in a graph, and are powerful enough to enable us to get rather quickly to fairly deep theorems. They are, however, not so useful for the study of linear dependence, so the connection between a "graphoid" and a matroid in the sense of Whitney (whose axioms are more appropriate for this purpose) is developed. The class of theorems we treat in this framework is essentially the well-known theorems, and some less well-known but highly important theorems, of graph-theory, electrical networks, and network-programming. (It is theorems of electrical-network *analysis* that are meant here, not *synthesis*. It is expected, however, that the ultimate contribution of matroid-theory to electrical theory will be in the synthesis area; it seems to the writer that there have been a number of ingenious schemes for network-synthesis which terminate with the discovery of an incidence-matrix which meets many of the requirements for being the "cut-set matrix" of a graph, and yet is not in general realizable as a graph. In such a case, it is very likely to be an incidence-matrix for a matroid. A better understanding of the nature of this pitfall would thus facilitate the development of better network-synthesis schemes.)

It may seem to the reader that the re-doing of theorems known in the context of graphs, electrical networks, etc., in this abstract context is of doubtful value, since it tends to render invisible that which was formerly visible. However, it has been the experience of Mathematics that this process of "axiomatization," "abstraction," or "formalization" is a very healthy thing for a subject—in the process of eliminating intuition in favor of logic, many interesting facets of the subject-matter are brought out clearly which were before hidden.

A purely formal development of this subject-matter would involve many trivial or repetitious proofs, or an excessive number of the usual excuses for not presenting them. The writer has taken the liberty of including them in the form of "Exercises" with perhaps over-explicit "Suggestions"; some less easy results have been included in this form in order not to break the thread of the main discussion.

As for preliminaries: It is assumed that the reader is familiar with the notions $A \cup B$, $A \cap B$, and $\bar{A}$ (complement of A) of the Boolean algebra of sets. The notation $A - B$ means $A \cap \bar{B}$, and $A \Delta B$ means $(A \cup B) - (A \cap B)$.

The notation $\{x\}$ means "the set whose only element is x"; we shall frequently write $A + x$, $A - x$ when more properly we should write $A \cup \{x\}$, $A - \{x\}$. When we want to indicate, in a Venn-diagram, a one-element set, we shall draw a "small" area. The empty set is called $\emptyset$.

$A \subset B$ ("A is contained in B") means $A \cap \bar{B}$ is the empty set. The phrase "A is properly contained in B" means $A \subset B$ but $A \neq B$. A set is called *minimal* with respect to a property p if it possesses property p but does not properly contain another set having property p. "Maximal" is defined similarly.

The reader is also assumed to be familiar with the notions of "field," "abstract vector space," "dimension," and "basis." (These concepts are now commonly taught in courses in "linear algebra for engineers" in the university, and can be found, e.g., in the book of Birkhoff and MacLane [3].)

The reader who is interested in *two-terminal networks* should shut this term out of his mind and concentrate on a *one-port network*, which is a *network with one line distinguished and treated differently from the others.* (The "distinguished line" can be thought of electrically as a piece of testing-apparatus attached to the two terminals.)

A *graph* consists of two finite sets (of undefined objects), L and P ("lines" and "points," or "elements" and "solder-joints," or "branches" and "nodes" or "edges" and "vertices") and a function assigning to each line an *unordered pair* of points. The *diagram* of a graph is a "picture" like that of Figure 1. It should be clear to the reader in what way the diagram "represents" the graph. Note that a "loop" is allowed (line with two end-points the same) because we said *unordered pair* and not "two-element set." A directed graph, or *digraph*, consists of lines and points and a function assigning to each line an *ordered* pair of points. Figure 2 shows the diagram of a digraph.

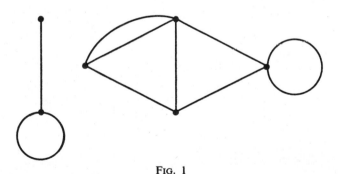

FIG. 1

In a graph: a *circuit* is a *set of lines* forming a simple closed curve. When we say "delete a line," we do *not* delete the associated points. A *cocircuit* is a *minimal separating set* of lines: a set whose deletion increases the number of connected components, but is minimal with respect to this property (does not *properly* contain another separating set).

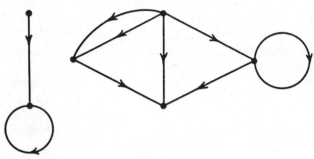

FIG. 2

In a *one-port graph* (a graph with a distinguished line *x*!) a *tie-set* is a circuit containing *x*, minus *x*; a *cut-set* is a cocircuit containing *x*, minus *x*. (These terminologies are not entirely standard; some authors use them for what we call "circuit" and "cocircuit.")

In the course of proofs, we have to make many "constructions." *No attempt is made to keep the constructions efficient.* There are two reasons for this: (1^0) it is not known at present what the important computational problems will be, especially the form in which the data will be presented; (2^0) it is usually more efficient to prove an existence-theorem without regard to the question of constructivity or efficiency of the proof, and afterwards to seek a construction, rather than search for a constructive or efficient proof.

2. GRAPHOIDS AND PRE-GRAPHOIDS.

In order to keep the definition of a "graphoid" concise, we introduce the notion of a *painting* of a finite set *L*. A painting of *L* is a partitioning of *L* into three subsets: *R*, *G*, *B*, such that *G* is a one-element set, i.e., $|G| = 1$. For easy visualization, it helps to think of the objects in R as being "painted red," the object in G as being "painted green," and the objects in B "painted blue." Note that $R \cap G$, $G \cap B$, and $R \cap B$ are all empty, and that $R \cup G \cup B = L$.

DEFINITION 2.1: A *graphoid* is a structure consisting of a finite set L of (undefined) objects, called "lines", and two collections **C**, **D** of nonempty subsets of L, called "circuits" and "cocircuits," satisfying the conditions:

(G-I) For any circuit C and cocircuit D: their intersection $C \cap D$ is not one line—i.e., $|C \cap D|$ may $= 0, 2, 3, 4$, etc., but $\neq 1$.

(G-II) For any painting of L: there exists either
 (i) a circuit C consisting of the green line and otherwise only red lines, or
 (ii) a cocircuit D consisting of the green line and otherwise only blue lines.

(G-III) No circuit contains another circuit properly; no cocircuit contains another cocircuit properly.

Axiom (G-II) is not particularly easy to visualize; the Venn-diagram of Figure 3 may help, but it is hard to draw a picture showing "either one thing or another" being true! Figure 3 also fails to illustrate the fact that we *did not require* that R be nonempty, or that B be nonempty.

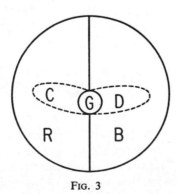

FIG. 3

EXERCISE 2.1: Show (using Axiom (G-I)) that the two possibilities of Axiom (G-II) are mutually exclusive (cannot occur simultaneously).

EXERCISE 2.2: Let m_1, m_2 be any two positive integers, with $m_1 + m_2 > 2$. Let L be a set consisting of $(m_1 + m_2 - 2)$ objects; let the "circuits" be the subsets of L consisting of precisely m_1 objects, and the "cocircuits" be the subsets consisting of precisely m_2 objects. Show that the structure is a graphoid. (Suggestion: for concreteness, try first $m_1 = 10$, $m_2 = 9$.)

The example of Exercise 2.2 is rather trivial, in the sense that later theorems will have nothing interesting to say about it. (It is, however, useful as a "counterexample," to show that some "theorems" about graphoids are not true.) A much more interesting example is as follows: take a finite linear graph **G**, and let L be the set of lines ("edges," or "branches") of **G**. Let a set of lines be called a "circuit" if they form a simple closed curve (note that the word "simple" excludes figure-eights!) and a "cocircuit" if they form a minimal separating set of lines. In this example, Axiom (G-I) holds because $|C \cap D|$ is always an *even* number, and (G-II) is a special case of a theorem of the writer [25, 26].

We caution the reader that *not all graphoids correspond to graphs.* (In fact, Exercise 2.2 exhibits a situation in which $|C \cap D|$ can be an odd number.) Thus pictures of graphs, although useful for the purpose of inspiring new theorems, can be highly misleading if one uses them to illustrate the proofs; the use of Venn-diagrams is highly recommended in preference.

DEFINITION 2.2: A pre-graphoid is a structure $(L, \mathbf{C}, \mathbf{D})$ satisfying (G-I) and (G-II), but not necessarily (G-III).

EXERCISE 2.3: In the context of Exercise 1: show that, if "circuit" means "set containing *at least* m_1 objects," and "cocircuit" means "set containing *at least* m_2 objects," we obtain a pre-graphoid. For what values of $(m_1 + m_2 - 2)$ does it fail to be a graphoid?

EXERCISE 2.4: Show that in a pre-graphoid the two alternatives of (G-II) are mutually exclusive.

Notice that every graphoid is automatically a pre-graphoid, so that *any theorem about pre-graphoids is a theorem about graphoids.*

DEFINITION 2.3: In a pre-graphoid: a line which is itself a circuit is called a *loop*; a line which is itself a cocircuit is called a *bridge*.

EXERCISE 2.5: Show that a line is a loop if and only if it is contained in no cocircuit. State the dual-theorem.

DEFINITION 2.4: In a pre-graphoid: If two lines form a circuit, they are said to be in *parallel*; if they form a cocircuit, they are said to be *in series*.

EXERCISE 2.6: Verify that in a graph, two lines need not have a common vertex (node) in order that they be "in series."

We now introduce the *incidence-matrices* of a graphoid (or pre-graphoid). Number the lines $1, \ldots, l$, the circuits $1, \ldots, c$, and the cocircuits $1, \ldots, d$. The *circuit incidence matrix* has entries c_{ij}, where $c_{ij} = 1$ if line j is in circuit i, and $c_{ij} = 0$ otherwise. The *cocircuit incidence matrix* is defined correspondingly. These matrices are shown in Figure 5 for the graphoid of the graph shown in Figure 4. We shall use the symbols $\mathbf{C}$, $\mathbf{D}$ to denote these matrices, even though we have used these symbols for another purpose. (No confusion will arise.) The *rows* of $\mathbf{C}$ may be denoted: $C_1, \ldots, C_c$, and those of $\mathbf{D} : D_1, \ldots, D_d$, again without danger of confusion. (Note that these rows are essentially the so-called "characteristic functions" of the corresponding sets.)

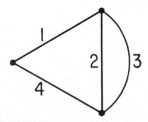

FIG. 4

$$\begin{bmatrix} 1 & 1 & 0 & 1 \\ 1 & 0 & 1 & 1 \\ 0 & 1 & 1 & 0 \end{bmatrix} \begin{bmatrix} 1 & 0 & 0 & 1 \\ 1 & 1 & 1 & 0 \\ 0 & 1 & 1 & 1 \end{bmatrix}$$

FIG. 5

3. TREES AND COTREES.

Throughout this section, we shall assume that we have before us a *graphoid* rather than a pre-graphoid (but see Exercise 4.4).

DEFINITION 3.1: A *tree* T is a set of lines which contains no circuit, and which is maximal with respect to this property (i.e.: augmenting T by adjoining any line produces a set which contains at least one circuit). A *cotree* is a set of lines containing no cocircuit, and maximal with respect to this property.

In graph-theory, the phrases "maximal tree," "spanning tree," "framework," and "skeleton" are often used. Our terminology "tree" follows the usage of the electrical engineers.

EXERCISE 3.1: What are the trees and cotrees of the graphoid of Exercise 2.2?

EXERCISE 3.2: For the graphoid of a graph: find a formula relating the number of lines in a tree, the number of vertices of the graph, and the number of connected components.

LEMMA 3.1: *Let T be a set of lines. Then T is a tree if and only if: T contains no circuit, and $\overline{T}$ (the set of lines not in T) contains no cocircuit.*

Proof: Suppose T is a tree. Now, if $\overline{T}$ contains a cocircuit D: let x be any line of D. Then $T + x$ contains a circuit C; this circuit contains x because T contains no circuit. But then $C \cap D$ is just x, and $|C \cap D| = 1$, contradicting Axiom (G-I). (Draw a Venn-diagram!)

Conversely: suppose T contains no circuit and $\overline{T}$ contains no cocircuit. We must show that for any line x in $\overline{T}$, $T + x$ contains a circuit. Paint all lines of T red, paint x green, and paint the remaining (unpainted) lines of $\overline{T}$ blue. (Draw a Venn-diagram!) Since there is no green-and-blue cocircuit, by Axiom (G-II) there is a green-and-red circuit.

EXERCISE 3.3: State the dual of Lemma 3.1, beginning with "Let S be a set of lines $\cdots$ "

THEOREM 3.1: *Let T be a tree. Then $\overline{T}$, the complement of T, is a cotree.*

Proof: By Lemma 3.1, T contains no circuit and $\overline{T}$ contains no cocircuit. The conclusion follows immediately by Exercise 3.3, putting $S = \overline{T}$.

THEOREM 3.2: *Let T be a tree, and x a line of $\overline{T}$. Then $T + x$ contains a unique circuit (which we shall name C_x).*

Proof: Suppose C_1 and C_2 are distinct circuits in $T + x$. Then either $C_1 - C_2$ or $C_2 - C_1$ is nonempty; without loss of generality, assume the former. (See Figure 6.)

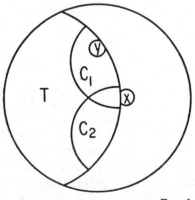

FIG. 6

Let y be a line of $C_1 - C_2$. Then, by Theorem 3.1, $\overline{T} + y$ contains a cocircuit D; since $|C_1 \cap D| \neq 1$, D contains x. But then $C_2 \cap D = x$, contradicting Axiom (G-I).

EXERCISE 3.4: State the duals of Theorem 3.1 and 3.2. (Notice that the dual of Theorem 3.2 is *not trivial* in the application to graph-theory!)

Now we come to a fairly deep theorem.

THEOREM 3.3: *Let* T_1, T_2 *be any two trees. Then* $|T_1| = |T_2|$.

EXERCISE 3.5: Check the truth of this theorem in the graphoid of Exercise 2.2.

Proof: Ignoring the trivial case $T_1 = T_2$, we see (by the "maximality" property of a tree) that neither contains the other, so $T_1 - T_2$ and $T_2 - T_1$ are nonempty. (See Figure 7.) Choose any line x of $T_2 - T_1$. Now, $T_1 + x$ contains a circuit C, which (since T_2 contains no circuit) must contain a line y of $T_1 - T_2$. (See Figure 8, where the heavy line delineates C.) Now, $\overline{T}_1$ is a cotree (Theorem 3.1) so $\overline{T}_1 + y$ contains a cocircuit D which in turn contains y; by Axiom (G-I) it must also contain x. (See Figure 9.)

By Theorem 3.2, $T_1 + x - y$ contains no circuit (because removing y from $T + x$ destroys the *unique* circuit C) and by Exercise 3.3, $\overline{T}_1 + y - x$ contains no cocircuit (for the exact dual-reason). It now follows from Lemma 3.1 that $T_1 + x - y$ is a tree.

Now (see Figure 10) we have a new tree T' which has the same number of lines as T_1 but is "one jump closer" to T_2. This tree may be exactly T_2; in this case the proof is complete. If not, we can repeat the above process, with T' in the place of T_1. Continuing in this way we get a sequence of trees $\{T_1, T', T'', \ldots, T^{(n)}\}$ such that each tree of the sequence has the same number of lines as its predecessor, and finally $T^{(n)} = T_2$.   Q.E.D.

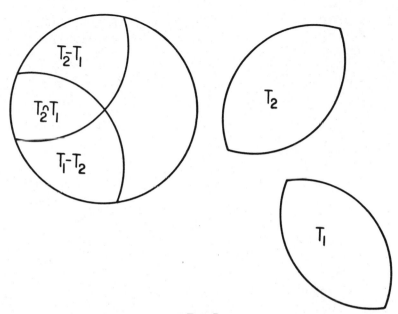

FIG. 7

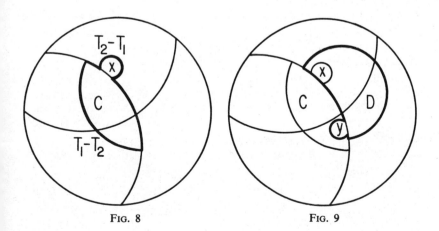

FIG. 8   FIG. 9

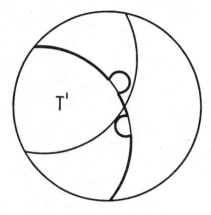

F<small>IG</small>. 10

The reader will recall that this is the same kind of process as is used in proving that two bases of a finite-dimensional vector space have the same number of vectors. We shall see that this connection is a deep (not superficial!) one in Exercise 5.10 later.

E<small>XERCISE</small> 3.6: State the dual of Theorem 3.3. Give a proof of this dual using Theorem 3.3, but which does *not* proceed by "dualizing" the theorem.

We now state a definition. We have no immediate use for the concept introduced here (i.e., we cannot at this point prove any interesting theorems involving the concept!) but this is the natural place to state it, anyway.

D<small>EFINITION</small> 3.2: Let T be a tree, and let $x_1, \ldots, x_s$ be the lines of $\bar{T}$. Then $(T + x_1), \ldots, (T + x_s)$ contain unique circuits $C_{x_1}, \ldots, C_{x_s}$. We call these the *fundamental system of circuits* associated with T, and shall sometimes call them simply $C_1, \ldots, C_s$.

E<small>XERCISE</small> 3.7: Formulate the dual definition to Definition 3.2.

E<small>XERCISE</small> 3.8: Let C be a circuit, and let x, y be two lines of C.

Show that there exists a cocircuit containing x and y but no other lines of C. (Suggestion: form a tree by adding lines to $(C - x)$.)

4. MORE ON PRE-GRAPHOIDS.

EXERCISE 4.1: Let (L, C, D) be a pre-graphoid. Let C' be any system of subsets of L such that (i) C' contains all of C, and (ii) each set C' in C' is a union of elements of C. Let D' be another system of subsets related to D in the same way as C' is related to C. Prove that (L, C', D') is a pre-graphoid.

Pre-graphoids are not very interesting objects *per se*; however, the concept is very useful in the study of graphoids. The theorems of this section are not very impressive, but we shall have good use for them later. From this point on, we shall use the terms "pre-circuit" and "pre-cocircuit" for the members of C and D of a pre-graphoid, for reasons which will be clear shortly. (See Definition 4.1.)

LEMMA 4.1: *In a pre-graphoid: suppose a pre-circuit C is not minimal—i.e., it contains properly another pre-circuit. Then C is a union of properly smaller pre-circuits.*

Proof: Let $C_1 \subset C, x \in C_1, y \in C - C_1$. (See Figure 11.) Paint y green, x blue, the rest of C red, and the rest of L blue (see Figure 12). Now if there is a pre-cocircuit D containing the green line and otherwise only blue lines, then to prevent $|C \cap D| = 1$ we must have $x \in D$. But then $|C_1 \cap D| = 1$. Thus there is no such D, and there is by (G-II) a pre-circuit C_y containing y but not x.

Now, let the lines in $C - C_1$ be called $y_1, y_2, \ldots, y_m$. Form a system of pre-circuits $C_{y_1}, \ldots, C_{y_m}$ as above, and observe that C is the union of these pre-circuits and C_1; these are all properly smaller than C.

LEMMA 4.2. *In a pre-graphoid: each pre-circuit is a union of minimal pre-circuits.*

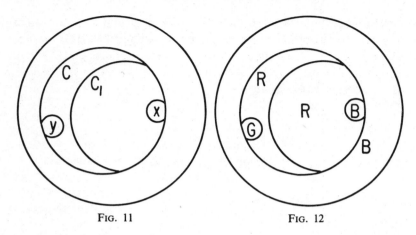

FIG. 11   FIG. 12

Proof: By Lemma 4.1, if a pre-circuit is not itself minimal, it can be broken up into (i.e., written as a union of) smaller pre-circuits. Each of these, in turn, if not itself minimal, can be broken up into smaller ones; and so on until we get all minimal pre-circuits.

Now we come to an important theorem:

THEOREM 4.1: *In a pre-graphoid* $(L, \mathbf{C}, \mathbf{D})$: *let* $\mathbf{C}'$ *be the collection of minimal pre-circuits, and* $\mathbf{D}'$ *the collection of minimal pre-cocircuits. Then* $(L, \mathbf{C}', \mathbf{D}')$ *is a graphoid.*

EXERCISE 4.2: Prove Theorem 4.1, using Lemma 4.2.

DEFINITION 4.1. In a pre-graphoid $(L, \mathbf{C}, \mathbf{D})$: the graphoid composed of all minimal pre-circuits and minimal pre-cocircuits is called the *underlying graphoid* of $(L, \mathbf{C}, \mathbf{D})$. These *minimal* pre-circuits and pre-cocircuits of the pre-graphoid will often be called its *circuits* and *cocircuits*.

The usefulness of Theorem 4.1 will be as follows: we shall sometimes want to construct a graphoid having certain properties. We can now proceed by constructing a pre-graphoid, and then by discarding the non-minimal precircuits and pre-cocircuits obtain the underlying graphoid, and then prove that it has the desired properties.

DEFINITION 4.2: In a pre-graphoid, we define a *tree* as a set of lines containing no pre-circuit, and maximal with respect to that property.

EXERCISE 4.3: What is the relationship between a tree of a pre-graphoid and a tree of the underlying graphoid?

EXERCISE 4.4: Observe that Axiom (G-III) was used in Section 3 in Ex. 3.8, so that all the results of that Section could have been proved for pre-graphoids as well as graphoids. Explain (in the light of the answer to Exercise 4.3) why it was not worthwhile to prove the theorems in this greater generality.

5.   CONNECTED GRAPHOIDS; GRAPHOIDS AND MATROIDS.

Let us introduce a *relation* S between the lines of a graphoid. We write $x_1 S x_2$ provided either $x_1 = x_2$ or there exists a circuit C containing x_1 and x_2.

EXERCISE 5.1: Verify that the relation S is "symmetric": $x_1 S x_2$ if and only if $x_2 S x_1$. Also, show that the word "cocircuit" could have been used instead of "circuit" in the above definition without changing the meaning of S. Suggestion: suppose $x_1 \neq x_2$, and both lines are in some circuit C. Paint x_1 green, x_2 blue, the rest of C red, and all remaining lines blue; apply (G-III), (G-II), and (G-I).

Now, let us write $x_1 \equiv x_2$ if either $x_1 S x_2$ or: there is a sequence $(x_1, x', x'', \ldots, x^{(n)}, x_2)$ of lines such that each line is S-related to its predecessor and successor in the sequence. (By the way, $(\equiv)$ is sometimes called the "transitive extension of S.")* It is easily seen that $(\equiv)$ is an equivalence-relation—i.e., it is reflexive, symmetric, and transitive. By a famous theorem (see [3]) the set L of lines can be partitioned into classes $L_1, L_2, \ldots, L_m$ such that: if x_1 and x_2 are in the same class, then $x_1 \equiv x_2$, and if they are in different classes, then $x_1 \not\equiv x_2$.

An equivalence-class of lines is called a *component* of the graphoid. If the graphoid has only one component, it is called *connected*; if more than one, it is called *separated* or *separable*.

EXERCISE 5.2: Show that if the components of a graphoid are $L_1, \ldots, L_m$, and $\mathbf{C}_1$ is the set of circuits containing lines of L_1, and $\mathbf{D}_1$ is the set of cocircuits containing lines of L_1: then L_1 together with $\mathbf{C}_1$ and $\mathbf{D}_1$ is a connected graphoid.

EXERCISE 5.3: What are the components of the graphoid of the graph of Figure 1? (There are *four* altogether, not *two*!)

EXERCISE 5.4: Suppose the lines of L are numbered so that the lines of L_1 are $1, \ldots, l_1$, the lines of L_2 are numbered $l_1 + 1, \ldots, l_1 + l_2$, and so on. With this kind of numbering, what can one say about the appearance of the line-circuit and line-cocircuit incidence matrices? (See Section 2.) Can one see the incidence-matrices of the graphoid of Exercise 5.2 in this picture?

DEFINITION 5.1: A matroid is a structure consisting of a finite set L of (undefined) objects called "lines" and collection $\mathbf{C}$ of

---

*It can be shown (see [41]) that $x_1 S x_2$ and $x_2 S x_3$ imply $x_1 S x_3$, and hence the transitive extension of S is S itself.

nonempty subsets of L, called "circuits", satisfying:

(M-I) No circuit contains another circuit properly.

(M-II) If C_1 and C_2 are any two circuits, $x \in C_1 \cap C_2$, $y \in (C_1 - C_2)$, then there exists a circuit C_3 such that $y \in C_3$, $x \notin C_3$, and $C_3 \subset (C_1 \cup C_2)$. (See Figure 13.)

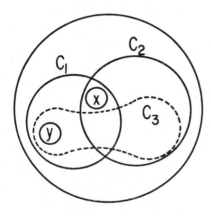

FIG. 13

EXERCISE 5.5: Let $(L, \mathbf{C}, \mathbf{D})$ be a graphoid. Show that $(L, \mathbf{C})$ is a matroid. (Suggestion: to show (M-II): paint y green, x blue, the rest of $C_1 \cup C_2$ red, and the rest of L blue. Use (G-I) and (G-II).) Also state the dual-theorem.

DEFINITION 5.2: Let $(L, \mathbf{C})$ be a matroid. Suppose there is a system $\mathbf{D}$ of subsets of L such that $(L, \mathbf{C}, \mathbf{D})$ is a graphoid. Then $(L, \mathbf{D})$ which is a matroid by Exercise 5.5, is called a *dual-matroid* of $(L, \mathbf{C})$. (Note: although this definition is the writer's, the original concept and another definition are due to Whitney [41].)

THEOREM 5.1: *Let $(L, \mathbf{C}, \mathbf{D})$ be a graphoid, and E a nonempty set of lines. Then E is a cocircuit if and only if: for any circuit C, $|C \cap E| \neq 1$, and E is minimal (nonempty) with respect to this property.*

Proof: Clearly if E is a cocircuit, $|C \cap E| \neq 1$ by (G-I). Also: suppose $E' \subset E$ but $E' \neq E$. Let $x \in E'$, $y \in (E - E')$ and apply Exercise 3.8 to conclude there exists a C such that $|E' \cap C| = 1$. Thus $E' = E$.

Now, assume E has the two properties, and let x be any line of E. Paint x green, the rest of E blue, and the rest of L red. (Draw a Venn-diagram!) By hypothesis on E, there is no green-and-red circuit, so by (G-II) there is a green-and-blue cocircuit D. $D \subset E$ and for any circuit C, $|C \cap D| \neq 1$. Thus if $E \neq D$, E cannot be minimal; hence $E = D$.

Let us give some interpretations of Theorem 5.1:

Interpretation 1. If we know the line-circuit incidence-matrix of a graphoid, then we can construct (reconstruct?) from it the line-cocircuit incidence-matrix.

Interpretation 2. If a matroid has a dual-matroid, then this dual-matroid is unique, so that we are justified in speaking of "*the* dual-matroid."

EXERCISE 5.6: Show that if a matroid **M** has **M′** as its dual, then **M′** has **M** as its dual.

Now we are finally going to make use of some of the theory of pre-graphoids which we developed earlier. We are leading up to the theorem that *every matroid has a dual-matroid.*

LEMMA 5.1: *Let* $(L, \mathbf{C})$ *be a matroid. Then there exists a system* **D** *of subsets of L such that* $(L, \mathbf{C}, \mathbf{D})$ *is a pre-graphoid.*

Proof: Let p be a painting of L such that there is no circuit consisting of the green line and otherwise only red lines. Form a set D_p as follows:

(i) D_p contains the green line.
(ii) For each circuit C containing the green line, D_p contains *at least* one blue line in common with C.
(iii) D_p is minimal with respect to properties (i) and (ii).

Notice that there is no doubt of the existence of a set satisfying (i) and (ii): one has only to take the set consisting of the green line and all those blue lines which are contained in circuits containing the green line. Notice also that for a given p, there may be several possible sets D_p; in this case, choose *one* of them arbitrarily and call it D_p.

Now, let **D** be the collection of all the sets D_p, with p ranging over all paintings of the type described above. (Note there are possible duplications.)

Obviously $(L, C, \mathbf{D})$ satisfies Axiom (G-II), and we need only show that it satisfies (G-I). Consider any circuit C and any set D_p, together with the associated painting p.

If C contains the green line, (G-I) is automatically satisfied, by property (ii) of D_p. Since D_p contains no red lines we can ignore the case where C consists of red lines only, and we have only to consider the case where C consists of blue and red lines (possibly none of them being red, but at least one being blue). Suppose $C \cap D$ is a single line x. We shall try to show a contradiction.

By property (iii) of D_p, there is a circuit C_1 which contains the green line, and otherwise has only the line x in common with D_p. Referring now to Figure 14, we see that the shaded area contains no line of D_p.

Now, by (M-II) there is a circuit C_2 containing the line G and otherwise entirely contained in the shaded area of Figure 14. But then it has only the line G in common with D_p, which contradicts either the definition of p or property (ii) of D_p. Q.E.D.

THEOREM 5.2: *For any matroid (L, C): there is a unique system* **D** *of subsets of L such that $(L, C, \mathbf{D})$ is a graphoid; i.e., every matroid has a unique dual-matroid.*

Proof: By Lemma 5.1, there is a pre-graphoid $(L, C, \mathbf{D}')$. Let $(L, C, \mathbf{D})$ be the underlying graphoid. (Definition 4.1.) (Note: it is easily seen from the way the underlying graphoid is formed that the circuits of the graphoid are the same as those of the matroid!) The uniqueness has already been stated and proved; see Interpretation 2 of Theorem 5.1.

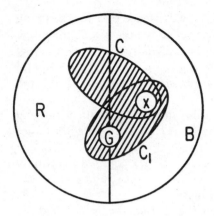

FIG. 14

EXERCISE 5.7. (optional) Is the pre-graphoid $(L, \mathbf{C}, \mathbf{D}')$ of the above proof actually a graphoid, so that the reduction to the underlying graphoid was really unnecessary?

EXERCISE 5.8: Let V be a vector space. Let L be a finite set of vectors in V. Define $\mathbf{C}$ as follows: C is in $\mathbf{C}$ if C is a (non-empty) *minimal linearly dependent set* of vectors of L. Prove that $(L, \mathbf{C})$ is a matroid.

Notice, now, that all the terminology we have for graphoids, like "tree," "cotree," "cocircuit," etc., have meaning for a matroid: they are taken as simply the corresponding concepts in the (unique) associated graphoid.

EXERCISE 5.9: For the matroid of Exercise 5.8: what is a *tree* of this matroid?

EXERCISE 5.10: Prove that any two bases of a finite-dimensional vector space have the same number of vectors. (Suggestion: let L be the union of the two bases, and use Exercise 5.9 and Theorem 3.3.)

One last comment before leaving this topic: G. Kron [21] has been heard to proclaim, not only that electrical networks have no nodes, but also that they have no cut-sets (cocircuits). One of the main points of this paper is that it is not only possible, but also desirable, to study graphs in such a way that nodes are never mentioned. However, it is clear from the content of this Section that it is possible, *but not desirable*, to dispense with the notion of cocircuit—for, given the circuit-matrix, one can always construct the cocircuit-matrix, and the cocircuits may be very useful objects to have at one's disposal as *tools*, even if one's primary interest is in the circuits.

6. NEW GRAPHOIDS FROM OLD.

(Later sections do not depend strongly on this section, so it could be omitted on first reading; the exercises are, however, rather valuable.)

For a graph **G**: two interesting operations, each of which produces a new graph **G'** (with one fewer line) are called *deleting a line* and *shrinking a line*. Concentrating on the latter, we see that when a line x is "shrunk to a point" (its two vertices then become one) the following phenomena occur:

(1^0) Any cocircuit containing x is "destroyed"—i.e., does not correspond to a cocircuit of **G'**. However, a cocircuit *not* containing x is *not* destroyed.

(2^0) Any circuit containing x goes over into a circuit with one line fewer.

(3^0) A circuit *not* containing x may cease to be a circuit, because it goes over into a union of *two* circuits.

Comment (3^0) seems to suggest that it will be easiest to mimic these considerations in graphoid-theory by dealing with a pre-graphoid rather than a graphoid.

Let $(L, \mathbf{C}, \mathbf{D})$ be a pre-graphoid with $|L| \geqslant 2$, and distinguish a line x. Let L' be $L - x$. For each pre-circuit C form a subset of L as follows: if $x \notin C$, let $C' = C$; if $x \in C$, let $C' = C - x$. (Note that if x is a loop, C' is the empty set.)

For each pre-cocircuit D not containing x, let $D' = D$.

Now let $\mathbf{C}'$ and $\mathbf{D}'$ be the collections of all the sets C' and D' formed as above (except that we do not use the empty set, if it appears as a C', in forming $\mathbf{C}'$).

EXERCISE 6.1: Show that if $C_1 \neq C_2$, then $C_1' \neq C_2'$; also, if $D_1 \neq D_2$, then $D_1' \neq D_2'$.

EXERCISE 6.2: Show that $(L', \mathbf{C}', \mathbf{D}')$ is a pre-graphoid.

EXERCISE 6.3: Show that if D is a cocircuit of $(L, \mathbf{C}, \mathbf{D})$—i.e., a minimal set in $\mathbf{D}$—then D' is a cocircuit in $(L', \mathbf{C}', \mathbf{D}')$.

DEFINITION 6.1: We shall say that $(L', \mathbf{C}', \mathbf{D}')$, formed as above from $(L, \mathbf{C}, \mathbf{D})$, is the pre-graphoid formed by *shrinking* x. If the roles of $\mathbf{C}$ and $\mathbf{D}$ are interchanged, it will be said to be formed by *deleting* x. The underlying graphoid of $(L', \mathbf{C}', \mathbf{D}')$ will be called *the graphoid obtained by shrinking x (resp. deleting x)*. We will ordinarily use this terminology only when $(L, \mathbf{C}, \mathbf{D})$ is a graphoid.

EXERCISE 6.4: For a pre-graphoid $(L, \mathbf{C}, \mathbf{D})$, show that if two lines x_1, x_2 are shrunk successively, the pre-graphoid $(L', \mathbf{C}', \mathbf{D}')$ obtained does not depend on the *order* in which they are shrunk. State the dual-theorem.

EXERCISE 6.5. Show that the pre-graphoid obtained by first shrinking x_1 and then deleting x_2 is the same as the one obtained by performing these operations in reverse order.

EXERCISE 6.6: Let P and Q be two disjoint subsets of L with $|P \cup Q| < |L|$. Show that it is meaningful to speak of "the pre-graphoid obtained by shrinking all lines of P and deleting all lines of Q"—i.e., that the order of shrinkings and deletions is immaterial. Suggestion: invent some notation!

EXERCISE 6.7: State what operations on the incidence-matrix correspond to the process of Exercise 6.6, and how to obtain the matrices of the underlying graphoid of the final pre-graphoid.

DEFINITION 6.2: We define a series-parallel (s.-p.) graphoid recursively as follows:

(i) A graphoid with one line is s.-p.
(ii) A graphoid is s.-p. provided either: (a) there are two lines in series (Definition 2.4) such that shrinking one of them produces an s.-p. graphoid, or (b) there are two lines in parallel such that deleting one of them produces an s.-p. graphoid.

EXERCISE 6.8: Show that any s.-p. graphoid is the graphoid of a graph, with C playing the role of the circuits and D the cocircuits of the graph. Show also that the dual-graphoid (L, D, C) is realizable as a graph!

EXERCISE 6.9: Show that if the line x is a loop, and two new graphoids are obtained by shrinking x and by deleting x respectively, these two new graphoids are identical.

EXERCISE 6.10: Show that if a loop is shrunk (or deleted) the trees of the graphoid are unchanged. What are the *cotrees* of the new graphoid (i.e., how are they related to the cotrees of the "old" graphoid)?

Now let us look at a more interesting way of making new graphs out of old ones. Two two-terminal graphs can be put together to form a new graph (with no distinguished terminals) in an obvious way—solder a terminal of the first to a terminal of the second, and then solder together the two "loose" terminals. (In mathematical terminology, "identify" is preferred to "solder together.") Can we mimic this procedure with graphoids? Recall that in this theory we shall have to deal with two *one-port* graphoids (L_1, C_1, D_1, x_1) and

$(L_2, \mathbf{C}_2, \mathbf{D}_2, x_2)$. We begin by letting $L' = L_1 \cup L_2 - x_1 - x_2$ (here L_1 and L_2 are assumed disjoint). A *circuit* of the new graphoid will be any one of the following: (i) a circuit of the first, *not* containing x_1; (ii) a circuit of the second, not containing x_2; (iii) for any circuit C_1 of the first which contains x_1, and any circuit C_2 of the second which contains x_2, form a circuit $C' = C_1 \cup C_2 - x_1 - x_2$. Form the *cocircuits* correspondingly.

EXERCISE 6.11: Show that the structure $(L', \mathbf{C}', \mathbf{D}')$ obtained as above is a graphoid.

EXERCISE 6.12: (For matroid-theorists only!) Take two copies of the complete graph on five points, distinguish a line of each, and think of them as one-port graphoids. Put them together as described above, but *reverse the roles of the circuits and cocircuits in one of the copies* before "assembly." Is the resulting graphoid realizable as the graphoid of a graph?

7. EVEN GRAPHOIDS AND BINARY MATROIDS.

DEFINITION 7.1: An even graphoid is a graphoid satisfying the following axiom: (G-I-E) for any circuit C and cocircuit D : $|C \cap D|$ is an even number.

It has already been remarked that the graphoid of a graph is an even graphoid. Thus everything we have to say about even graphoids will automatically be applicable to graph-theory. Notice that we now need pay no attention to Axiom (G-I), since it is implied by (G-I-E).

EXERCISE 7.1: Show that the graphoid of Exercise 2.2 is not (in general) an even graphoid, so that the conjecture "every graphoid is even" is in fact false.

DEFINITION 7.2: An even pre-graphoid is a pre-graphoid whose pre-circuits and pre-cocircuits satisfy (G-I-E).

EXERCISE 7.2: Let (L, C, D) be an even pre-graphoid. Let C' be related to C as follows: C' contains all of C; any remaining set of C' is a *disjoint union* of sets in C—i.e., it can be decomposed into disjoint elements of C. Let D' be related similarly to D. Show that (L, C', D') is an even pre-graphoid.

Even pre-graphoids (like other pre-graphoids!) are not very interesting objects *per se*, but are useful tools in the development of the theory of even graphoids.

EXERCISE 7.3: Let (L, C', D') be an even pre-graphoid, and let (L, C, D) be the underlying graphoid. Show that (L, C, D) is an even graphoid.

Let us investigate further the relationship between an even pre-graphoid and its underlying graphoid.

THEOREM 7.1: (*Context as in Exercise* 7.3.) *Any pre-circuit* $C' \in C'$ *is a disjoint union of circuits. Furthermore, for any circuit* C_1 *properly contained in* C', *the disjoint decomposition can be chosen so as to include* C_1.

Proof: Suppose C' is not a circuit. Then there is a circuit C_1 properly contained in C; let x_1 be a line of C' not in C_1. (See Figure 15.) Color x_1 green, C_1 blue, the rest of C' red, and the rest of L blue. Suppose there is a cocircuit D containing the green line and otherwise only blue lines. Then clearly either $|C \cap D|$ or $|C_1 \cap D|$ is odd; both these possibilities are forbidden. Hence, since the underlying graphoid obeys Axiom (G-II), there is a red-and-green circuit C_2.

Now, C_1 and C_2 are disjoint (see Figure 16). Either they fill up C' or there is a line x_2 left over. Color x_2 green, C_1 and C_2 blue, the rest of L blue (Figure 17). If there is a cocircuit D containing

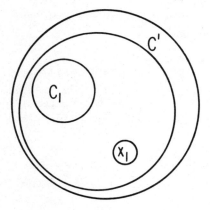

FIG. 15

the green line and otherwise only blue lines, then at least *one* of $|C' \cap D|$, $|C_1 \cap D|$, $|C_2 \cap D|$ must be odd, since the former is the sum of the latter two, plus 1; all these are forbidden, so there is a green-and-red circuit C_3.

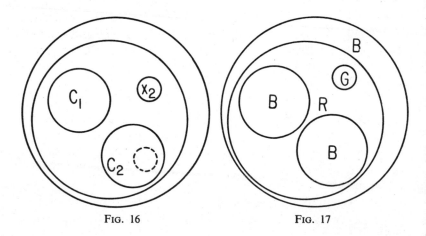

FIG. 16   FIG. 17

The argument continues along these lines until C' is exhausted: $C' = C_1 \cup C_2 \cup \cdots \cup C_m$, with all C_i disjoint from each other.

EXERCISE 7.4: Rewrite the above proof in a formal way (without using the phrase "and so on" or any of its disguises).

DEFINITION 7.3: A *binary matroid* [41] is a matroid with the property: For any tree T and circuit C: let $x_1, \ldots, x_m$ be the lines of C which are not in T. Then $C = C_1 \Delta C_2 \Delta \cdots \Delta C_m$, where Δ is "symmetric difference" and C_i is the fundamental circuit corresponding to T and x_i. (See Definition 3.2).

The primary object of this section is to elucidate the connection between even graphoids and binary matroids.

EXERCISE 7.5: Let $(L, \mathbf{C})$ be a binary matroid, $(L, \mathbf{D})$ its dual. Show that $(L, \mathbf{C}, \mathbf{D})$ is an even graphoid. Suggestions: for any cocircuit D, form a cotree containing all but one line of D, and let T be the associated tree. Write the circuit C in terms of the fundamental-circuits corresponding to T.

EXERCISE 7.6: (For graph-theorists only!) Show that the circuit-matroid of a graph is binary; show also that the cocircuit-matroid is binary. (This exercise will be given again later, when we have enough tools to make it easy.)

LEMMA 7.1: *Let* $(L, \mathbf{C}, \mathbf{D})$ *be an even pre-graphoid. Let* C_1, C_2 *be two distinct pre-circuits (with* $|C_1 \Delta C_2| \neq 0$). *Form* $\mathbf{C}'$ *as* $\mathbf{C}$ *with* $C_1 \Delta C_2$ *adjoined. Then* $(L, \mathbf{C}', \mathbf{D})$ *is an even pre-graphoid. Furthermore, the underlying (even) graphoid of* $(L, \mathbf{C}', \mathbf{D})$ *is the same as that of* $(L, \mathbf{C}, \mathbf{D})$.

EXERCISE 7.7: Prove Lemma 7.1, following this outline:

(1^0) Show $(L, \mathbf{C}', \mathbf{D})$ satisfies Axiom (G-II).
(2^0) Show it satisfies Axiom (G-I-E). (Draw a Venn-diagram!)
(3^0) Show that $(L$, minimal elements of $\mathbf{C})$ and $(L$, minimal elements of $\mathbf{C}')$ are matroids having the same dual-matroid, and are hence the same (see Exercise 5.6).

LEMMA 7.2: *Let $(L, \mathbf{C}, \mathbf{D})$ be an even pre-graphoid. Let $\mathbf{C}'$ be composed of $\mathbf{C}$ and some nonempty symmetric differences (like $C_1 \Delta C_2 \Delta \cdots \Delta C_m$) of elements of $\mathbf{C}$. Then $(L, \mathbf{C}', \mathbf{D})$ is an even pre-graphoid whose underlying (even) graphoid is the same as that of $(L, \mathbf{C}, \mathbf{D})$.*

Proof: Consider first the case where $\mathbf{C}'$ consists of *all* the nonempty symmetric differences. Then $\mathbf{C}'$ can be built up by adjoining "single" symmetric differences (like $C_1 \Delta C_2$) one-at-a-time, and Lemma 7.1 can be applied in each step.

EXERCISE 7.8: Complete the proof of Lemma 7.2 for the case where some of the nonempty symmetric differences are "missing."

LEMMA 7.3: *Let $(L, \mathbf{C}, \mathbf{D})$ be an even pre-graphoid. Let $\mathbf{C}'$ consist of $\mathbf{C}$ plus some certain nonempty symmetric differences of elements of $\mathbf{C}$, and $\mathbf{D}'$ be formed analogously from $\mathbf{D}$. Then $(L, \mathbf{C}', \mathbf{D}')$ is an even pre-graphoid.*

EXERCISE 7.9: Prove Lemma 7.3 as follows: first use Lemma 7.2, then use the dual of Lemma 7.2, and complete the proof following the outline of Exercise 7.8.

LEMMA 7.4: *Let $(L, \mathbf{C}, \mathbf{D})$ be an even graphoid, and T a tree. Let x and y be distinct lines of $\overline{T}$. Let C be a circuit containing x and y, and otherwise only elements of T. Let C_1 and C_2 be the (unique!) circuits containing x and y respectively, and otherwise only elements of T. Then $C = (C_1 \Delta C_2)$.*

Proof: We have to show (A) that every line of C is a line of $C_1 \Delta C_2$, and (B) that every line of $C_1 \Delta C_2$ is a line of C.

Proof of (A): Those lines of C which are in $\overline{T}$ are obviously in $C_1 \Delta C_2$. If we can show that no line of C is in $C_1 \cap C_2$, and no line of C is in $T - (C_1 \cup C_2)$, we will be through. (See Figure 18.)

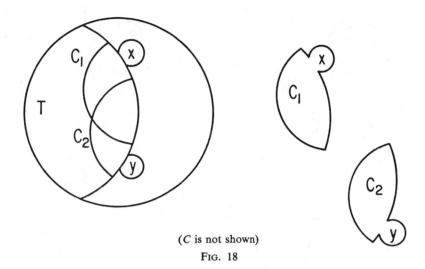

(C is not shown)

FIG. 18

Suppose $z \in C$ and $z \in (C_1 \cap C_2)$. Since $\overline{T}$ is a cotree, we can form the fundamental cocircuit D consisting of z and otherwise only lines of $\overline{T}$. By Axiom (G-I) applied to C_1 and C_2, x and y are lines of D (see Figure 19). But then $C \cap D$ must consist of the lines x, y, and z, contradicting Axiom (G-I-E).

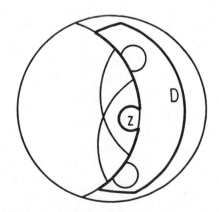

(C is not shown)

FIG. 19

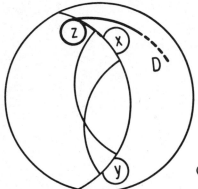

(*C* is not shown; *D* only partially shown)

FIG. 20

Now suppose $z \in C$, $z \in T - (C_1 \cup C_2)$. Form the fundamental cocircuit D just as above (see Figure 20). Apply Axiom (G-I) to C and D to see that D *must* contain x or y, so that either $|D \cap C_1| = 1$ or $|D \cap C_2| = 1$, in violation of Axiom (G-I). (This paragraph is referred to in Exercise 7.10 below.)

We have finished the proof of (A), and move on to (B). Let $\mathbf{C}'$ be formed by adjoining $C_1 \Delta C_2$ to $\mathbf{C}$. By Lemma 7.1 above, $(L, \mathbf{C}', \mathbf{D})$ is a pre-graphoid, and its underlying graphoid is $(L, \mathbf{C}, \mathbf{D})$. By Theorem 7.1, $C_1 \Delta C_2$ is a disjoint union of circuits, and the decomposition can be chosen to include C. (Note this is true because we proved (A) above!) but if there were any other circuit, say C_1, in the decomposition, then $C_1 \subset T$ (draw a Venn-diagram!) which is impossible. Hence $C = C_1 \Delta C_2$.

EXERCISE 7.10: One of the paragraphs of this proof uses only Axiom (G-I), not (G-I-E). What theorem about graphoids (not necessarily even graphoids) is really proved in that paragraph?

We now come to a highly important theorem.

THEOREM 7.2: *The two matroids of an even graphoid are both binary matroids.*

Proof: We shall show that $(L, \mathbf{C})$ is a binary matroid. Consider the even graphoid $(L, \mathbf{C}, \mathbf{D})$, let T be a tree and C be a circuit; let

$C - T$ be the lines $x_1, \ldots, x_m$, let $\overline{T} - C$ be the lines $x_{m+1}, \ldots, x_s$, and let C_i be the (unique) circuit contained in $T + x_i$. We shall proceed by mathematical induction on m; note that the case $m = 1$ is obvious, and the case $m = 2$ is covered by Lemma 7.4.

Form the tree T' by adjoining x_m to T and then removing a line y of $(C_m \cap T) - C$ (just as in the proof of Theorem 3.3). By the induction-hypothesis, since C has only the lines $x_1, \ldots, x_{m-1}$ outside of T', we can write $C = C_1' \Delta C_2' \Delta \cdots \Delta C_{m-1}'$ (the meaning of these symbols is obvious). We now show that each C_i' is a symmetric difference of fundamental circuits associated with T. It is easy to see that C_i' has either one or two lines outside of T; in the former case, it is itself a fundamental circuit of T, and the latter case is covered by Lemma 7.4.

We now know that C is a symmetric difference of fundamental-circuits associated with T. Is it precisely *that one* referred to in the Theorem? It must be, for: using the rules for manipulating symmetric differences $E \Delta E = \emptyset$, $\emptyset \Delta E = E$, the expression can be reduced to one with no repeated terms; and if any C_i with $i \neq 1, \ldots, m$ appears, then the expression would contain a line outside of C; also, each C_i with $i = 1, \ldots, m$ obviously appears.

Having finished the proof of this formidable theorem, we see that we have proved a highly nontrivial theorem of graph-theory:

EXERCISE 7.11: *Now* do Exercise 7.6, using Theorem 7.2.

A few words are now in order concerning digital computation (although we generally stay away from that topic in this paper). Consider a relatively small graph—say, the complete graph on 8 points. The numbers of circuits and cocircuits of this graph are so enormous that corestorage of the circuit-and cocircuit-incidence matrices is virtually unthinkable (with present-day machines). This suggests that even for "rather small" matroids, (having, say, on the order of 28 lines) it might be very difficult to store the matroid-structure in the machine. But this is not so for binary matroids! One only has to know one tree and the associated fundamental-circuits; this small amount of information contains the complete matroid-structure. For: the collection of all nonempty symmetric differences of these fundamental-circuits contains all the circuits and otherwise only symmetric-differences of circuits. Lemma 7.2 then asserts that by rejecting all these sets which are not minimal, we obtain all the circuits of the matroid. This procedure is not computationally feasible for a moderate-sized matroid, but we shall ignore the problem of giving a computationally feasible scheme, simply because it is not known to the writer at present whether this will turn out to be an important problem. It is only the question of *storage*-feasibility which we are attempting to answer here.

A rather compact, and well-known to electrical network-theorists, method for storing the structure (for binary matroids) is as follows: number (or renumber, if necessary!) the lines of a cotree $1, \ldots, s$, and suppose L has l lines in all. Form the $s \times l$ incidence-matrix of the fundamental circuits. This matrix will have the partitioned form $[I \,|\, A]$, where I stands for the identity-matrix. The matrix A then contains (implicitly) the complete matroid-structure.

But the most important feature of binary matroids (and even graphoids) is of an algebraic character.

DEFINITION 7.4: Let **C** and **D** be the incidence-matrices of a graphoid with l lines. Consider the vector space (over the field F of integers mod 2, whose elements are 0 and 1) of l-tuples of zeros and ones, called F^l. The *circuit-space mod* 2 of the graphoid (which could also be called "current-space mod 2") is the subspace of F^l generated by the rows of **C**; the *cocircuit space mod* 2 is the subspace generated by the rows of **D**. A *current* (*mod* 2) is an element of the current-space mod 2; a *voltage* (*mod* 2) is an element of the cocircuit-space mod 2. (In the remainder of this Section, we shall drop the phrase "mod 2"; it is understood throughout.) These definitions "make sense" in a general (non-even) graphoid, but are not especially interesting in that context.

DEFINITION 7.5: Given a tree T: a *fundamental-circuit vector* is an n-tuple of 0's and 1's having its 1's in precisely those entries corresponding to a fundamental-circuit (with respect to *T*). A *fundamental-cocircuit-vector* is defined analogously.

THEOREM 7.3: *Consider an even graphoid and consider any tree T. Then the fundamental-circuit-vectors are a basis for the circuit-space; the dual theorem also holds.*

Proof: It is obvious that these vectors are linearly independent, since each has a 1 in a position where all the others have 0's; thus no one of these vectors is a linear combination of the others.

Now, Theorem 7.2 tells us that any circuit *C* can be written as a symmetric difference of fundamental-circuits. It is routine to check that if we add up the corresponding vectors (mod 2) we get the vector corresponding to *C*. Thus any sum of vectors corresponding to *C*'s can be written in terms of the fundamental-circuit-vectors.

(This phenomenon is known to mathematicians as *isomorphism*; the correspondence between vectors and the corresponding subsets of *L* is an isomorphism carrying the operation + into Δ. Incidentally, this isomorphism and the well-known theorem which states that a vector can have *only one* representation as a linear combination of a set of independent vectors, is an alternative proof for the last paragraph of the proof of Theorem 7.2.)

The reader who wishes to know more about binary matroids should now refer to Whitney's original paper [41]; if he wishes to see contact with more concrete subject-matter, he can now refer to Lehman's paper [23] on matroids vis-à-vis switching theory. If, on the other hand, he wants to try out his wings as a matroid-theorist, I would suggest that he try to put Duffin's paper on the Wang algebra of networks [9] into matroid-language!

8. DIGRAPHOIDS, ELECTRICAL NETWORKS,
   AND NETWORK-PROGRAMMING.

We now wish to consider what part of the theory of *directed* graphs can be built up in the wider context of graphoids/ matroids. To this end, we shall introduce the concepts of *digraphoid* (short for "directed graphoid") and *orientable graphoid*. Now, Tutte [38] has implicitly defined an orientable matroid as one on which a regular Abelian chain-group can be built up, and it is (essentially) oriented by choosing one particular such chain group. However, this work has a very formidable and technical appearance, and this notion of "orientation" has very little direct intuitive appeal. Although the objects we shall study are essentially a regular Abelian chain-group and its dual (axiomatized simultaneously), our approach will involve much more elementary "primitive" notions than Tutte's. The connection with regular chain-groups is discussed in the Appendix.

It is necessary to introduce a notion of orientation into electrical-network theory for the treatment of non-bilateral elements, such as rectifiers. The notion of an *orientable* graphoid is a valuable one because some of the theorems (e.g., on networks of bilateral elements) do not require the distinguishing of any particular orientation for their *statement*—the orientation is needed as a *tool in the proof.*

Let us briefly review and classify "methods of electrical-network analysis." The writer believes that four classes of theory are sufficient:

I. The so-called "topological" method. This method was originated by the great graph-theorist Kirchhoff (see, e.g., [19]) and advanced by Feussner [12], [13]. These fundamental works are for some reason not well-known to engineers; the basic ideas were recently rediscovered by Percival [34], [35].

II. The "linear algebra and matrix-theory" method. This has its origins in the "mesh-analysis" method of J. Clerk Maxwell, and is computationally so convenient that electrical engineers often think of it as "the" method. It has probably retarded the study of nonlinear electrical networks in the same way that the Laplace transform has retarded the study of nonlinear differential equations. It does, however, lend itself well to the study of networks containing triodes operated in the linear range.

III. The "variational" method. The beginnings of this method are due to Maxwell, who proved the first (known to the writer) extremum-properties of solutions of electrical networks. Duffin [8] became aware of the extendibility of these principles to nonlinear networks, and used them to prove existence-theorems; the same ideas were rediscovered and stated explicitly by Millar [24] and popularized by Cherry. A synthesis with methods of solving variational problems of the Operations Research field was seen by Charnes and Cooper [6]. The theory was further developed by the writer [25] and a lucid treatment is given by Berge [2]. General "pseudo-variational" methods for the treatment of a large class of nonlinear problems of mathematical physics are given by the writer [29], [27], [32], F. E. Browder [4], [5], and Zarantonello [42], [43] but have been very little applied to network-analysis because of their newness.

IV. The "digital topological" method. This was pioneered by Ford and Fulkerson (see [14]), but in a way which does not bring out its relevance to either electrical networks or matroid/graphoid theory. A complete treatment, in a form making the theory immediately transferable to the context of digraphoids, is given in the writer's works [25], [26]. A good treatment (in which, however, the fundamental existence theorems for solutions are missing) is given by Berge [2], using the writer's algorithm. Appendix B of the present paper contains the most important of the results.

But let us begin. We shall call the structures we are about to introduce "digraphoids" (short for "directed graphoids"), in analogy with the term "digraph" (for "directed graph"). We present an inscrutable-looking axiom-system first and explain it later.

DEFINITION 8.1: A *digraphoid* is a structure consisting of: (1^0) a graphoid, and (2^0) a *partitioning* of each circuit and cocircuit of the graphoid, *each* being partitioned into two sets; this partitioning is to satisfy the axiom:

(DG) For any circuit C and cocircuit D: let C', C'' and D', D'' be the partitioning (note $|C' \cap C''| = 0$, $C' \cup C'' = C$, etc.). Then $|C' \cap D'| + |C'' \cap D''| = |C' \cap D''| + |C'' \cap D'|$.

Notice that it was *not* prescribed that both of C', C'' (or both of D', D'') be nonempty.

EXERCISE 8.1: Show that, in Axiom (DG), it *does not matter* which of the two sets of the partitioning of C, is called C' and which C''.

Let us now re-state the definition in more understandable but less concise form.

DEFINITION 8.2: A graphoid is called *orientable*, or *directable*, if it is possible to change some of the 1's in the incidence-matrices C, D to (-1)'s in such a way that each row of C (corresponding to a *circuit*) is *orthogonal* to each row of D (corresponding to a *cocircuit*). Here, the numbers $+1$, -1, 0 are to be treated as real numbers (or integers) in computing the "dot-product" (or "inner product"). (Notice that we are abusing notation by referring to the *signed* matrices as C and D.)

To make an example: Figure 21 shows the incidence-matrices C and D of an orientable graphoid; Figure 22 shows C and D after signs have been introduced, proving that it is indeed orientable.

$$
\begin{bmatrix} 1 & 0 & 0 & 0 \\ 0 & 1 & 1 & 1 \end{bmatrix}
\qquad
\begin{bmatrix} 0 & 1 & 1 & 0 \\ 0 & 0 & 1 & 1 \\ 0 & 1 & 0 & 1 \end{bmatrix}
$$

FIG. 21

$$
\begin{bmatrix} +1 & 0 & 0 & 0 \\ 0 & +1 & -1 & +1 \end{bmatrix}
\qquad
\begin{bmatrix} 0 & +1 & +1 & 0 \\ 0 & 0 & -1 & -1 \\ 0 & -1 & 0 & +1 \end{bmatrix}
$$

FIG. 22

EXERCISE 8.2: Show that any orientable graphoid is an even graphoid.

EXERCISE 8.3: Show that, given an establishment of signs as in Definition 8.2, the resulting partitioning of each circuit and co-circuit into a "plus-set" and a "minus-set" satisfies Axiom (DG).

EXERCISE 8.4: Show that the graphoid of any digraphoid is orientable. (Use the partitionings!)

Having done Exercise 8.4, the reader will notice that, *for each row* of the matrices **C**, **D**, it *does not matter* which of (say) C', C'' is taken as the plus-set and which the minus-set. When signs have been established as in Exercise 8.4, we shall call the resulting matrices *the incidence-matrices of the digraphoid*, noticing that we are abusing the definite article "the" by using it to describe a *non-unique* pair of objects. Definition 8.1 avoids this abuse, because it does *not* say that each C and D is partitioned into an *ordered* pair of subsets. Otherwise expressed: any set of rows of "the" matrices **C**, **D** can be multiplied through by (-1) without changing the digraphoid represented by the matrices.

From now on, we shall abuse the word "the," *in this connection only*.

EXERCISE 8.5: Verify that the graphoid of any graph is orientable. (Suggestion: "draw arrows on the lines" of the graph!)

DEFINITION 8.3: Let S be a set of lines of a digraphoid. Multiplying through by (-1) *all columns corresponding to lines of S* in both **C** and **D** produces a new digraphoid, called the "digraphoid obtained by reversing the orientations of the lines of S." (Note that we *do not define the concept* "orientation of a line," however!)

EXERCISE 8.6: Show that the structure produced by the operation of the above definition is indeed again a digraphoid.

Now let G be any Abelian group. (The reader who is unfamiliar with the definition may substitute: "let G stand for any of (i) the integers mod 2, (ii) the integers, (iii) the real numbers, or (iv) the complex numbers." This substitution is inadequate, but will serve as a crutch.) We shall call G *the coefficient-group*. (There is no abuse of "the" here, since although G might be any one of, say, four things, we are developing four theories "in parallel," and in any one of these theories, G is just one single object!)

DEFINITION 8.4: Suppose we take the circuit-incidence-matrix C of a digraphoid; replace every (1) in row 1 by an element g_1 of G; every (1) in row 2 by an element g_2 (possibly the same as g_1); and so on. (The minus-signs remain where they are.) Then add the rows to produce an n-tuple of elements of G. Any n-tuple which can be produced in this way is called a "current," or a "flow." For a single fixed G: the set of *all* currents is called "the current-space (relative to G)."

We define "voltage-drop" and "voltage-drop space" analogously, with D in place of C. (The phrase suggests that we should previously have defined "voltage"; however, this concept has no place in the present theory.) The current-space will be called K', and the voltage-drop-space K''. The letter is chosen in honor of Kirchhoff. (Perhaps a topologist would prefer P' and P'', since they correspond to familiar objects in homology-theory.)

EXERCISE 8.7: In case G is a field F (say, the real or complex numbers) show that K' and K'' are vector-spaces over F.

EXERCISE 8.8: (Important!) In case G is the real or complex numbers, show that K' and K'' are orthogonal complements in R^l (resp. C^l). Suggestion: let d', d'' be their dimensions. Let t be the number of lines in a tree, s the number of lines in a cotree. Show as in Theorem 7.3 that $d' \geqslant s$ and $d'' \geqslant t$. Then observe that every vector of K' is orthogonal to every vector of K'', so that $d' + d'' \leqslant l$, by the theorem "the sum of the dimensions of two subspaces is the dimension of the sum plus the dimension of the intersection."

EXERCISE 8.9: (For functional analysts only!) Let H be a Hilbert space, with real or complex scalars, and take $G = H$. Show that K' and K'' are orthogonal complements in H^l.

We remark that Exercise 8.8 contains implicitly the famous formula

$$\sum_{i=1}^{l} e_i i_i = 0,$$

and Exercise 8.9 contains the formula

$$\sum_{i=1}^{l} \int_{-\infty}^{+\infty} e_i(t) i_i(t) dt = 0.$$

Both formulas, of course, are conservation-of-energy principles. (In the latter case, consider $H = L^2(-\infty, +\infty)$.)
We now formulate some problems of electrical-network theory.

PROBLEM 1. For given (fixed) complex numbers $z_1, \ldots, z_l$ and $E_1, \ldots, E_l$, does there exist a $2 \times l$ matrix of complex numbers

$$\begin{bmatrix} i_1, \ldots, i_l \\ e_1, \ldots, e_l \end{bmatrix}$$

such that $(i_1, \ldots, i_l)$ is in the current-space, $(e_1, \ldots, e_l)$ is in the voltage-drop space, and: for each column, $e_j = i_j z_j + E_j$ $(j = 1, \ldots, l)$? If so, give formulas for the solution.

PROBLEM 2. Same as Problem 1, but with the real numbers in place of the complex numbers throughout.

We present the outline of the scheme called "mesh-analysis" for the solution of these problems. First: the dimension of the current-space is known (from Exercise 8.8) to be s, the number of lines in a cotree. Let $c_1, \ldots, c_s$ be linearly independent vectors in K' (in "mesh-analysis," they are usually chosen for convenience to be linearly independent rows of C, obtained for example as in Exercise 8.8.) Let $x_1, \ldots, x_s$ be "unknown" coefficients. Then set $x_1 \mathbf{c}_1 + \cdots + x_s \mathbf{c}_s$ = the first row of the unknown matrix. This vector can be written as

$$(x_1 c_{11} + \cdots + x_s c_{s1}, \ldots, x_1 c_{1l} + \cdots + x_s c_{sl}).$$

Set

$$e_i = (x_1 c_{1i} + \cdots + x_s c_{si}) z_i + E_i, \quad \text{for} \quad i = 1, \ldots, l.$$

The way we demand that the vector $\mathbf{e}$ be in K'' is to demand that it be orthogonal to K', or equivalently, that it be orthogonal to a basis for K'. So let $\mathbf{c}'_1, \ldots, \mathbf{c}'_s$ be a (possibly different!) set of independent rows of C. We write the equations:

$$\sum_{j=1}^{l} c'_{mj} e_j = \sum_{j=1}^{l} c'_{mj} \left[z_j \left(\sum_{k=1}^{s} x_k c_{kj} \right) + E_j \right] = 0.$$

This is now a system of s equations in s unknowns x_k. After solving for these, we can construct the solution (the $2 \times l$ matrix) in an obvious way.

It is clear that Problem 1 (or Problem 2) has a solution *if and only if* the set of simultaneous equations has a solution. We cannot guarantee that the problem has a solution without further hypotheses on the numbers z_j and/or E_j, however.

EXERCISE 8.10. Write out in full the dual of the above process, using $i_j = e_j a_j + I_i$ as "given" equations (here, a_j and I_j are assumed "known").

EXERCISE 8.11: Rewrite the mesh-analysis process with maximum use of matrix notation; specialize it to the case where $\mathbf{c}'_j = \mathbf{c}_j$ and the $\mathbf{c}_j$ are derived from a tree as in Exercise 8.8; write the matrices $[c_{ij}]$ in partitioned form, with the "identity-matrix" displayed explicitly where it occurs.

Exercise 8.10 brings us close to the method called "node-analysis"; however, we cannot approach closer because we have no concept of "node" in digraphoid-theory!

EXERCISE 8.12: Develop all of Section 6 over again for digraphoids—i.e., show how to construct "new digraphoids from old" in each of the ways suggested there; be sure to prove that the resulting structures *are*, indeed, digraphoids. Suggestion: draw pictures of digraphs for inspiration, but be careful not to use them in the proofs!

This is as far as we shall carry the "linear algebra and matrix theory" approach to electrical network analysis; i.e., just far enough to demonstrate its feasibility in the digraphoid-context.

We shall not delve at all into the "topological" method. The writer is confident that all the classical formulas for the solution of Problems 1 and 2 can be developed in the digraphoid-context—I shall leave it as a "research problem."

Let us turn now to the "digital topological" method. To show the feasibility of doing network-analysis in this style in the digraphoid-context, it is sufficient to prove the fundamental "Lemme des Arcs Colorés," as it has been called by Berge [2]. (The original theorem, for graphs, is due to the writer [25]; an expository presentation of the theorem is given in Reference 26.) We first introduce the notion of a *painting* of a digraphoid. This consists of a partitioning of the lines into three sets, R, G, and B, and the distinguishing of one line of the set G. One can think of it as "a painting of the lines with three colors, each line being red, green, or blue, and exactly one green line being colored dark-green."

THEOREM 8.1: ("*Lemme des Arcs Colorés*" *for digraphoids.*) *Given a digraphoid; for any painting of the lines* (*as defined above*) *there exists one, but not both, of:*

(i) *A circuit containing the dark-green line but no blue lines, in*

> which all the green lines are similarly-oriented (*i.e.*, *have all
> the same sign in the incidence-matrix* **C**), *or*
> (ii) *A cocircuit containing the dark-green line but no red lines, in
> which all the green lines are similarly-oriented.*

Proof: The proof will proceed by mathematical induction on the number of green lines. If there is only one (the dark-green line) the conclusion follows by Axiom (G-II).

Now suppose the theorem has been proved when there are m green lines, and consider the case of $(m + 1)$ green lines. Choose a green line x other than the dark-green line (see Figure 23).

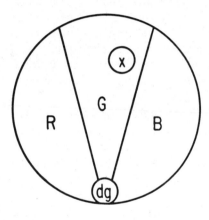

Fig. 23

Color the line x red. If there is a cocircuit of type (ii), we are through.

Now color x blue. If there is a circuit of type (i), we are through.

Suppose neither of these phenomena occurs. Then, by the induction-hypothesis, there is a cocircuit of type (ii) when x is painted blue, and a circuit of type (i) when x is painted red. Let us examine the corresponding rows of the incidence-matrices (Figure 24). In this figure, we have assumed the lines numbered so that the dark-green line comes first, then the red lines, then the blue lines, then the remaining green lines, and finally x. We have also chosen the sign-conventions so $(+1)$ appears in the "dark-green" position of both vectors.

$$
\begin{array}{c|ccccccc|ccccc|ccccc|c}
dg & & & R & & & & & & B & & & & G & & & x \\
\hline
D: (+1 & 0 & 0 & 0 & 0 & 0 & 0 & +1 & 0 & -1 & +1 & 0 & +1 & +1 & 0 & ?) \\
C: (+1 & 0 & +1 & -1 & 0 & -1 & -1 & 0 & 0 & 0 & 0 & 0 & +1 & 0 & +1 & ?)
\end{array}
$$

FIG. 24

By Axiom (DG), these two vectors are orthogonal. The contribution to the inner product from the dark-green line is $+1$; from all the red and blue lines, zero; from the green lines, a nonnegative integer p; and from x, an unknown number q, which must be 0, $+1$, or -1. Now, we have $1 + p + q = 0$, so $1 + q \leqslant 0$, and obviously $q = -1$. Thus one of the question-marks in Figure 24 is $+1$, and the other is -1. Choosing the vector in which it is $+1$, we have the desired circuit or cocircuit.

EXERCISE 8.13: In Theorem 8.1: show that both (i) and (ii) cannot be present. Suggestion: use the orthogonality of the two corresponding rows of the incidence-matrices.

The development of the writer's papers [25, 26], which give a complete theory of nonlinear networks of two-terminal elements with "monotonic" current-vs.-voltage-drop characteristics, will now go through essentially without difficulty. (There are two arguments concerning the finiteness of certain constructions which must be modified to transfinite constructions, but this is an easy matter—see Appendix B.) Since the fundamental existence-, uniqueness-, and extremum-principle theorems for such networks *are also* the fundamental theorems of the field of Operations Research mathematics called "network-programming," it is clear that digraphoid-theory also forms a natural context for network-programming. A review of the most fundamental theorems is given here in Appendix B.

The special case of *linear* networks does not require so many definitions. For example, we can state:

THEOREM 8.2: *In Problem 2 above: if all* $z_j \geqslant 0$, *and if there is no circuit such that* $z_j = 0$ *for all lines of the circuit, then the equations of the mesh-analysis method have a solution, and the solution is unique.*

Proof: Follows immediately from the theorems of Reference 25, thought of in the digraphoid-context. (The existence-theorem is given as Theorem B3 of our Appendix B, taking $E_j = \{(i, e):$ $e = z_j i + E_j\}$.)

Let us give an existence-uniqueness theorem for Problem 1. In the proof, I shall violate my promise to use no tools but elementary linear algebra.

THEOREM 8.3: *In Problem 1 above: if all* z_j *lie in the right-hand part of the complex plane, then the equations of the mesh-analysis method have a unique solution.*

Proof: It is easy to find a real positive constant c such that for all j, $\mathrm{Re}\, z_j \geqslant c$. Now, consider the linear transformation A on C^l (the space of l-tuples of complex numbers) defined by: $A(i_1, \ldots, i_l) = (i_1 z_1, \ldots, i_l z_l)$. Let us compute the real part of the inner product $\langle A\mathbf{i}, \mathbf{i} \rangle$; it is

$$\mathrm{Re}\langle A\mathbf{i}, \mathbf{i} \rangle = \mathrm{Re} \sum_{j=1}^{l} z_j i_j \bar{i}_j$$

$$= \tfrac{1}{2} \sum_{j=1}^{l} z_j i_j \bar{i}_j + \sum_{j=1}^{l} \bar{z}_j \bar{i}_j i_j$$

$$= \sum_{j=1}^{l} \frac{z_j + \bar{z}_j}{2} |i_j|^2$$

$$= \sum_{j} (\mathrm{Re}\, z_j)|i_j|^2.$$

It follows that

$$\mathrm{Re}\langle A\mathbf{i}, \mathbf{i}\rangle \geqslant c\|\mathbf{i}\|^2.$$

Now, let A' be the restriction of A to K', and let P be the orthogonal projection-operator on K'. We have, for $\mathbf{i} \in K'$,

$$\mathrm{Re}\langle PA'\mathbf{i}, \mathbf{i}\rangle \geqslant c\|\mathbf{i}\|^2,$$

and by a famous theorem on bounded linear operators in Hilbert space ("the closure of the numerical range contains the spectrum"), we see that 0 is not in the spectrum of PA', considered as an operator on K'. Consequently, the equation

$$0\mathbf{i} + PA'\mathbf{i} = PE$$

has a unique solution in K'. Thus $A'\mathbf{i} - \mathbf{E} \in K''$, or since $\mathbf{i} \in K'$, we can write $A\mathbf{i} - \mathbf{E} \in K''$. The proof is complete.

9. CONCLUSION AND ACKNOWLEDGEMENT.

The writer hopes that one of the purposes of this paper has been accomplished, and that the reader previously unacquainted with matroid-theory has become sufficiently interested in it to dig more deeply into the subject via the already-published literature. We have been able to present only a sampling (fairly representative, to be sure) of the kinds of theorems easily provable in this context rather than the more restricted context of graph-theory.

The writer wishes to acknowledge his debt to Professor W. T. Tutte for reading several earlier versions of this paper and pointing out to what extent their contributions were original (and to what extent not!) The writer regrets that proper acknowledgement of the work of H. Whitney in the usual way—by crediting specific theorems—has not been possible because of the reorganization of the subject-matter on the basis of new axiom-systems; let it suffice to say that his influence is all-pervasive in this paper.

Appendix A

Digraphoids and Regular Chain-Groups.

Tutte [38] makes the following definitions (we shall completely change the terminology but not the concepts). Given a finite set L of undefined objects (called "lines"), numbered (for convenience) $1, \ldots, n$, a *chain* is an n-tuple of integers $(x_1, \ldots, x_n)$. The *support* of a chain is the subset of L for which the corresponding entries of the chain are nonzero. A *chain-group* is a set A of chains (on a fixed set L) such that, for any two chains x_1, x_2 in L, the difference

$$\mathbf{x}_1 - \mathbf{x}_2 = (x_1' - x_2', \ldots, x_1^n - x_2^n),$$

is in A. It follows automatically that the zero-chain, the negative of any chain in G, and the sum of any two chains in G, are all in G.

A chain $\mathbf{x}$ is called *elementary* if there is no (nonzero) chain with smaller support than $\mathbf{x}$. If *in addition*, $x_1, \ldots, x_n$ are all ± 1 or 0, the chain is called *primitive*.

Now, the chain-group A is called *regular* if to every elementary chain in G, there corresponds a primitive chain with the same support. The elements (n-tuples) of A will in this case be called *integer-valued current flows*.

Tutte now defines the *dual* A^* of a regular chain-group A as the set of all chains orthogonal to it. He shows (his number (5.1)) that the dual is regular and that $(A^*)^* = A$.

Now, let us call the supports of elementary chains in A, *circuits*, and the supports of elementary chains in A^*, *cocircuits*. Let us define "tree" and "cotree" just as we did for graphoids. Tutte shows (his number (5.2)) that the complement of a tree is a cotree, and (of course) vice-versa.

Let us call the set of circuits $\mathbf{C}$, and the set of cocircuits $\mathbf{D}$, and show that $(L, \mathbf{C}, \mathbf{D})$ is an even graphoid.

Axiom (G-I-E) follows from the definition of the dual (orthogonality). Axiom (G-III) comes from the definition of "elementary chain." Now consider any painting of L. If there is no red-and-green circuit, then by repainting some (possibly none) blue lines so that they become red, we can form a green-and-red tree. By the theorem of Tutte referred to above, the blue lines are now a cotree, so there is a cocircuit containing the green line and otherwise only blue lines; this proves Axiom (G-II)

It now follows immediately from Tutte's definition of "dual" that the *primitive* chains of A and A^* yield a digraphoid.

Conversely: let us now show that the circuit-space of a digraphoid, using the integers as coefficients, is a regular chain-group. Choose any elementary chain in the circuit-space. It might, for example, have the form $x = (0, 0, + 5, - 5, + 5, 0, 0, + 5)$. However, we do not *know* yet that the nonzero entries all have the same absolute value. Paint *any* line of its support green, the rest of its support red, and the rest of L, blue. If there were a green-and-blue cocircuit, the corresponding row of the (oriented) cocircuit-matrix could not be orthogonal to x; hence, since Axiom (G-II) holds, there is a green-and-red circuit. The corresponding row of the circuit-matrix is a primitive chain whose support is contained in that of x; hence (since x is elementary) is the same as that of x.

Thus we have demonstrated that the structures studied under the name of "digraphoids" are really the primitive chains of a regular chain-group and its dual. However, we believe that the digraphoid-axioms will be more palatable to the engineer than those of Tutte.

Appendix B.

Theorems of Network-Programming.

This material is placed in an appendix because it has a rather technical appearance, and is not essential if the reader's interest is in digraphoids as a basis for the theory of *linear* electrical networks.

Theorem B1: *Given a digraphoid (or equivalently, by Appendix A, a regular chain-group) with lines numbered $1, \ldots, n$. Let G stand for either of the real numbers or the integers. Let K' be the current-space. (If G is the integers and we are in the context of a regular chain-group, K' is the regular chain-group.)*

Let intervals in $G : I_1, \ldots, I_n$ be given. Does there exist an element $(i_1, \ldots, i_n)$ of K' such that, for each $j = 1, \ldots, n$, we have $i_j \in I_i$?

The answer is "yes" if and only if: for each oriented cocircuit (row of the oriented-cocircuit matrix) we have

$$0 \in \sum_{j=1}^{n} \epsilon_j I_j$$

where: the ϵ_j are the entries of this row; $(-1)I_j$ means the set of all negatives of elements of I_j; $(+1)I_j = I_j$; $(0)I_j$ is 0; and the sum of sets is the set of all sums of representatives.

Proof: Discussion postponed.

THEOREM B2: (*Max-flow-min-cut theorem for digraphoids.*)

Let a one-port digraphoid (or regular chain-group) be given, with lines numbered $1, \ldots, n$, the distinguished line being numbered 1; G is the reals or the integers. Let constants $c_j \geqslant 0$ (for $j = 2, \ldots, n$) be given. An admissible flow is defined as a vector $(i_1, \ldots, i_n) \in K'$ such that $-c_j \leqslant i_j \leqslant c_j$ for $j = 2, \ldots, n$. What is the maximum value that i_1 may have in an admissible flow?

The answer is the minimum, over the collection of all cut-sets (see§ 1) of

$$\sum c_j,$$

where only the c_j corresponding to the lines of the cut-set appear in the sum.

Proof: Follows easily from Theorem A1, by letting b be a real number, putting $I_1 = \{b\}$, putting $I_j = [-c_j, c_j]$ (the closed interval bounded by $-c_j, c_j$) for $j = 2, \ldots, n$, and inquiring for the maximum value which b can have so that there is an admissible flow.

THEOREM B3: (*A fundamental theorem on existence of solution of a nonlinear electrical network.*)

Let G be the reals or the integers. In $G \times G$, write (i_1, e_1) $M(i_2, e_2)$ provided $(i_1 - i_2) \cdot (e_1 - e_2) \geq 0$. Define a resistor as a subset E of $G \times G$ such that: for any (i_1, e_1) and (i_2, e_2) in G, $(i_1, e_1)M(i_2, e_2)$ and E is maximal with respect to this property. (*If G is the reals, this is seen to be "a curve going upward and to the right."*)

Now let a digraphoid (or a regular chain-group and its dual) be given, with lines numbered $1, \ldots, n$. Let K', K'' be the current-space and voltage-drop space over G. (*If G is the integers, and we are in the regular chain-group context, K' is the chain-group, and K'' is the dual.*) Let resistors $E_1, \ldots, E_n$ be given.

Does there exist a pair of vectors $(i_1, \ldots, i_n)$ and $(e_1, \ldots, e_n)$ with entries in G, with $(i_j, e_j) \in E_j$ for $j = 1, \ldots, n$?

The answer is "yes" if and only if: the projections of the E_j on the i-axes satisfy the conditions of Theorem B_1, and the projections on the e-axes satisfy the same conditions, but with "cocircuit" replaced by "circuit."

Proofs of Theorems B1 and B3: These proofs are word-for-word the same as the proofs of Theorems 4.1, 7.3, and 8.1 of [25]. Lemma 7.2 (*i*) (and all that follows from it) is deleted, and the finite construction is replaced by an existence-theorem and easy transfinite (contrapositive) proof. Before attempting to read these proofs (or concurrently) the reader should probably read [26], which is an expository discussion of the existence-proofs.

Remark. The engineer will prefer to replace the term "resistor" in the above theorem by the phrase "characteristic of a monotone nonlinear resistor."

THEOREM B4: *Let a digraphoid and resistors $E_1, \ldots, E_n$ be given; let indefinite integrals $F_j(i)$ and $G_j(e)$ be defined as in [25], pp. 200–201. Suppose the system has a "solution" in the sense of*

Theorem B3. *Then* $(i_1, \ldots, i_n)$ *solves the* "*programming problem*":

$$\min \sum_{j=1}^{n} F_j(i_j),$$

$$subj. \ to \ (i_1, \ldots, i_n) \in K',$$

and $(e_1, \ldots, e_n)$ *solves the dual-problem*

$$\min \sum_{j=1}^{n} G_j(e_j),$$

$$subj. \ to \ (e_1, \ldots, e_n) \in K''.$$

The proof is identical with that of the Corollary on p. 203 of [25]. In the above programming problems, it is understood that the variables are *also* constrained by the requirement that the F_j and G_j *are defined* for the values under consideration.

In mathematical programming, these extremum-problems are solved following the recipe of [28], which uses the algorithm of [26] and Theorem B4 above. In electrical problems, it is Theorem B3 which is considered as important, B4 being only a "bonus."

Let us give *one* theorem about "orientable" graphoids, the *line-form of Menger's Theorem* (see [11] for its statement in the context of graphs).

THEOREM B5: *Given an orientable one-port graphoid. Let a collection S of paths be called admissible if, for any two paths in the collection, no line of p_1 is a line of p_2. Then*

$$\max|S| = \min|D|$$

where S ranges through all admissible collections, and D ranges through all cut-sets (see § 1).

Outline of proof. It is clear *even without the orientability* that max $|S| \leqslant$ min $|D|$; the proof is left to the reader.

To show equality we must produce an S and a D for which $|S| = |D|$.

Number the distinguished line ("port") 1, and the rest 2, ..., n. Let $c_j = +1$ for $j = 2, \ldots, n$. Using Theorem B2, with G as the integers, we see that there exists an admissible flow $(i_1, \ldots, i_n)$ with $-1 \leqslant i_j \leqslant +1$ for $j = 2, \ldots, n$ and with

$$i_1 = \sum_D c_j = |D|,$$

for some D. We now produce S as follows. (*) Let line 1 be painted green; the jth line is blue if $i_j = 0$, green if $i_j = \pm 1$, for $j = 2, \ldots, n$. Reorient lines so that all i_j and $\geqslant 0$ for $j = 1, \ldots, n$. Apply the *Lemme des Arcs Colorés* (Theorem 8.1 of this paper.) Now, there can be no blue-and-green cocircuit with all green lines similarly oriented, for then this row of the cocircuit-matrix would not be orthogonal to $(i_1, \ldots, i_n)$. Thus there is a green circuit with all lines similarly oriented. This circuit contains a path (rejecting line 1). Subtract this row of the circuit-matrix from $(i_1, \ldots, i_n)$ with $i'_1 = i_1 - 1$. Repeat this process from (*).

The above construction is easily followed if the orientable graphoid corresponds to a graph. It is easily seen that repetition of the process (*) produces a sequence S of line-disjoint paths with $|S| = i_1$, but we know $i_1 = |D|$.   Q.E.D.

(**Note by author, 1973**): In retrospect, I feel that the principal defect or omission of this paper is its failure to exhibit that orientable graphoid/matroid theory and the theory of regular chain groups are intimately connected with (in fact, essentially synonymous with) the theory of totally unimodular matrices. As a beginning exercise in this direction, one can try proving that the "nullspace" of a totally unimodular matrix is a regular chain group. (Only vectors with integer entries are admitted to the "nullspace".)

REFERENCES

1. Berge, C., *Théorie des Graphes et ses Applications*, Paris, 1958. English translation New York: Wiley, 1962.

2. Berge, C., and A. Ghouila-Houri, *Programmes, Jeux et Réseaux de Transport*, Paris: Dunod, 1962.

3. Birkhoff, G., and S. Mac Lane, *A Survey of Modern Algebra*, New York: Macmillan, 1953.

4. Browder, F. E., "On the Solvability of Non Linear Functional Equations," *Duke Math. J.*, **30** (1963) 557–566.

5. ———, "Variational Boundary Value Problems for Quasi-Linear Elliptic Equations, II," *Proc. Nat. Acad. Sciences*, **50** (1963), 592–598.

6. Charnes, A., and W. W. Cooper, "Nonlinear Network Flows and Convex Programming over Incidence-Matrices," *J. Math. Anal. Appl.*, **5** (1962), 200–215.

7. Dolph, C. L., and G. J. Minty, "On Nonlinear Integral Equations of the Hammerstein Type," *Nonlinear Integral Equations*, Wisconsin: 1964.

8. Duffin, R. J., "Nonlinear Networks, IIa", *Bull. Amer. Math. Soc.*, **53** (1947), 963–971.

9. ———, "An Analysis of the Wang Algebra of Networks," *Trans. Amer. Math. Soc.*, **93** (1959), 114–131.

10. ———, "The Extremal Length of a Network," *J. Math. Anal. Appl.*, **5** (1962), 200–215.

11. Elias, P., A. Feinstein and C. E. Shannon, "Note on Maximum Flow Through a Network," *I.R.E. Trans. on Information Theory*, *IT-2* (1956), 117–119.

12. Feussner, W., "Uber Stromverzweigung in netzförmigen Leitern," *Ann. Physik*, **9** (1902), 1304–1309.

13. ———, "Zur Berechnung der Stromstärke in netzförmigen Leitern," *ibid.*, **15** (1904), 385–394.

14. Ford, L. R., Jr., and D. R. Fulkerson, *Flows in Networks*, Princeton: 1962.

15. Gallai, T., "Maximum-Minimum Sätze über Graphen," *Acta Math. Acad. Sci. Hungar.*, **9** (1958), 395–434.

16. Harary, F., "Graph Theory and Electrical Networks," *IRE Trans. on Circuit Theory*, **CT-6** (1959), 95–109.

17. Heller, I., and C. B. Tompkins, "An Extension of a Theorem of Dantzig," in Kuhn, H., and A. Tucker, *Linear Inequalities and Related Systems, Ann. of Math. Study*, **38**, Princeton: 1956.

18. Iri, M., "Comparison of Matroid Theory with Algebraic Topology with Special Reference to Applications to Network Theory," *RAAG Res. Notes*, University of Tokyo, **83** (1964).

19. Kirchhoff, G., "On the Solution of the Equations from the Investigation of the Linear Distribution of Galvanic Currents," *IRE Trans. on Circuit Theory*," **CT-5** (1958), 4–7.

20. König, D., *Theorie der endlichen und unendlichen Graphen*, New York: Chelsea, 1950.

21. Kron, G., "Graphs as Illegitimate Models of Electrical Networks," *Matrix Tensor Quart.*, **12** (1961), 1–10.

22. Lehman, A., "A Solution to the Shannon Switching Game," *Proc. IRE*, **49** (1961), 1339.

23. ——, *ibid., SIAM J.*, **12** (1964), 687–725.

24. Millar, W., "Some General Theorems for Nonlinear Systems Possessing Resistance," *Phil. Mag.*, **42** (1951), 1150–1160.

25. Minty, G. J., "Monotone Networks," *Proc. Roy. Soc., Ser. A*, **257** (1960), 194–212.

26. ——, "Solving Steady-State Nonlinear Networks of 'Monotone' Elements," *IRE Trans. on Circuit Theory*, **CT-8** (1961), 99–104.

27. ——, "Monotone (Nonlinear) Operators in Hilbert Space," *Duke Math. J.*, **29** (1962), 341–346.

28. ——, "On an Algorithm for Solving Some Network-Programming Problems," *Operations Res.*, **10** (1962), 403–405.

29. ——, "Two Theorems on Nonlinear Functional Equations in Hilbert Space," *Bull. Amer. Math. Soc.*, **69** (1963), 691–692.

30. ——, "On Axiomatic Foundations of the Theories of Directed Linear Graphs, Electrical Networks, and Network-Programming," *Summary of Papers of the 7th Midwest Symposium on Circuit Theory*, Ann Arbor, Mich.: 1964.

31. ——, "On the Monotonicity of the Gradient of a Convex Function,"

Pacific J. Math., **14** (1964), 243–247.

32. ——, "A Theorem on Maximal Monotonic Sets in Hilbert Space," *J. Math. Anal. Appl.*, **11** (1965), 434–439.

33. Ore, O., "Theory of Graphs," *Amer. Math. Soc Colloq. Publ.*, **38** (1962).

34. Percival, W. S., "The Solution of Passive Electrical Networks by Means of Mathematical Trees," *Proc. IEEE*, **100** (1953), 143–150.

35. ——, "Improved Matrix and Determinant Methods for Solving Networks," *ibid.*, **101** (1954), 258–265.

36. Rota, G.-C., "On the Foundations of Combinatorial Theory, I— Theory of Moebius Functions," *Zeitsch. für Wahrscheinlichkeitsrechnung und Verw. Gebiete.*, **2** (1964), 340–360.

37. Seshu, S., and M. B. Reed, *Linear Graphs and Electrical Networks*, Reading, Mass.: Addison-Wesley, 1961.

38. Tutte, W. T., "A Class of Abelian Groups," *Canad. J. Math.*, **8** (1956), 13–28.

39. ——, "Matroids and Graphs," *Amer. Math. Soc., Transl.*, **90** (1959), 527–552.

40. ——, "An Algorithm for Determining Whether a Given Binary Matroid is Graphic," *Proc. Amer. Math. Soc.*, **11** (1960), 905–917.

41. Whitney, H., "On the Abstract Properties of Linear Dependence," *Amer. J. Math.*, **57** (1935), 509–533.

42. Zarantonello, E. H., "The Closure of the Numerical Range Contains the Spectrum," *Bull. Amer. Math. Soc.*, **70** (1964), 781–787.

43. ——, "The Closure of the Numerical Range Contains the Spectrum," *Univ. of Kansas Technical Report*, **7** (1964).

HAMILTONIAN CIRCUITS

C. St. J. A. Nash-Williams

1. INTRODUCTION

A *circuit* is a non-empty finite connected graph in which the valency (or degree) of each vertex is 2 (Fig. 1). A *Hamiltonian circuit* of a finite graph G is a circuit, contained in G, which includes all the vertices of G. For example, the graph in Fig. 2 has a Hamiltonian circuit whose edges are indicated by thick lines in the figure, but the graphs in Figs. 3 and 4 have no Hamiltonian circuits, as the reader can probably convince himself by a little experimentation.

A persistent theme in graph theory has been a desire to determine, in some reasonable sense, which graphs have Hamiltonian circuits and which have not, i.e., we want necessary and sufficient conditions for a graph to have a Hamiltonian circuit. Of course, such necessary and sufficient conditions must be of a psychologically satisfactory kind, and we should not, for example, want a theorem which merely said, perhaps in a slightly disguised form, that a graph has a Hamiltonian circuit if and only if it has a Hamiltonian circuit. Probably the desired theorem would say that,

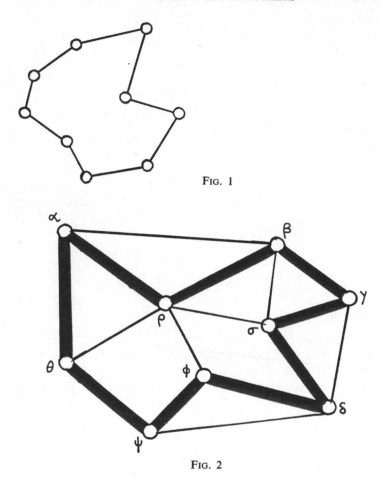

Fig. 1

Fig. 2

for every graph G, either G has a Hamiltonian circuit or the "shape" or "structure" of G has some particular feature which fairly obviously precludes the presence of a Hamiltonian circuit, and when we realize that this obstruction to the existence of a Hamiltonian circuit must rule out apparently promising graphs like those of Figs. 3 and 4, the difficulty of finding (let alone proving) the right conjecture can be recognized. Indeed, it may be that, in the very nature of things, no such necessary and sufficient

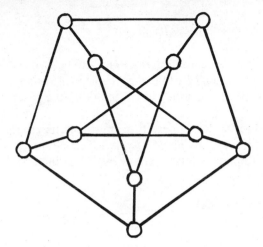

FIG. 3

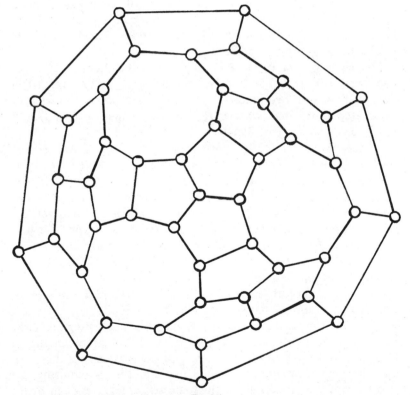

FIG. 4

condition for a graph to have a Hamiltonian circuit exists. Even if it exists, however, experience suggests that the problem of discovering it might well be of the same order of difficulty as the Four Colour Problem.

This situation has, however, not deterred graph-theorists from studying the problem and obtaining some results which, although far from constituting a complete solution, are nevertheless interesting. This paper will review some of these.

2. HAMILTONIAN CIRCUITS AND VALENCIES

We shall sometimes call a graph *round* if it has a Hamiltonian circuit and *tortuous* if not. (Many authors call a graph a *Hamiltonian graph* if it has a Hamiltonian circuit, but I personally prefer to use a different adjective.)

The *valency* of a vertex ξ of a graph is the number of edges incident with ξ. It will usually be denoted by $v(\xi)$. However, when we are discussing two graphs G, H and ξ is a common vertex of these graphs, we shall write $v_G(\xi)$ for the valency of ξ in G and $v_H(\xi)$ for the valency of ξ in H.

In the remainder of this paper, all graphs considered will be understood to be *simple*, i.e., without loops or multiple edges. This does not significantly limit the generality of our discussion: in a non-simple graph with at least three vertices, removal of all loops and all but one of the edges joining each pair of adjacent vertices will not affect the roundness or tortuosity of the graph. Furthermore, throughout this paper, all graphs which we consider will be understood to be *finite* simple graphs and the word "graph" will mean "finite simple graph."

Since we cannot at present find necessary and sufficient conditions for a graph to have a Hamiltonian circuit, we might ask whether we can at least find any reasonably interesting *sufficient* conditions. It seems reasonable to guess that a graph might be certain to have a Hamiltonian circuit if, in some sense, there are enough edges present, and hence that conditions requiring the valencies of the vertices to be, in some sense, large enough might guarantee the presence of a Hamiltonian circuit. A succession of results of this nature were proved by G. A. Dirac [7], L. Pósa [19], J. A. Bondy [1] and V. Chvátal [5], in that chronological order. Each of these results (after the first) strengthened the preceding one. After some preliminary definitions and lemmas, we shall present a proof of the first and weakest of these results (Theorem 1 below), and then some further definitions and lemmas will enable us to prove the last and strongest one (Theorem 2).

DEFINITIONS: In this paper, the letter G will always denote a graph. The set of vertices of G will be denoted by $V(G)$ and the set of edges of G will be denoted by $E(G)$. If two vertices ξ and η are joined by an edge, we shall say that ξ is *adjacent* to η and write ξ adj η. If G is a subgraph of a graph H, we shall say that H is a *supergraph* of G. If G is a subgraph of H and $V(G) = V(H)$, we shall say that G is a *spanning subgraph* of H and H is a *spanned supergraph* of G. A *one-edge extension* of G is a graph H obtained from G by adding an edge joining two vertices which are not adjacent in G. In other words, a one-edge extension of G is a spanned supergraph of G which has exactly one more edge than G. A graph G will be called *hypertortuous* if G is tortuous but every one-edge extension of G is round (i.e., if G has no Hamiltonian circuit but no further edge can be added to G without giving it a Hamiltonian circuit).

If $\lambda \in E(G)$, the graph obtained from G by removing the edge λ will be denoted by $G - \lambda$: thus $G - \lambda$ is a spanning subgraph of G. Somewhat analogously, we may sometimes use the symbol $G + \mu$ to denote a one-edge extension of G obtained by adding an edge μ joining two vertices which are non-adjacent in G.

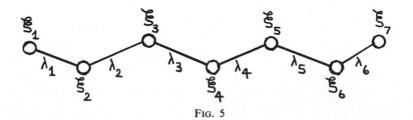

FIG. 5

An *ordering* of a finite set S is a sequence $x_1, x_2, \ldots, x_n$ obtained by arranging the elements of S in a definite order: for instance, the sequence 2, 8, 9, 3, 1, 5, 4, 6, 7 is an ordering of the set of positive integers less than 10. A graph P is a *path* if there is an ordering $\xi_1, \xi_2, \ldots, \xi_n$ of $V(P)$ and an ordering $\lambda_1, \lambda_2, \ldots, \lambda_{n-1}$ of $E(P)$ such that λ_i joins ξ_i to ξ_{i+1} for $i = 1, 2, \ldots, n - 1$: Fig. 5 illustrates this definition for $n = 7$. [A graph with one vertex and no edge counts as a path, since, in the foregoing definition, we may allow n to be 1 and "$\lambda_1, \lambda_2, \ldots, \lambda_{n-1}$" to be the empty sequence. But the "empty" graph, which has no vertices and no edges, is not considered to be a path.] A *Hamiltonian path* of G is a path in G which includes all the vertices of G, i.e., a path which is a spanning subgraph of G. If P is a path with at least two vertices, then the two vertices of P which are each incident with only one edge of P will be called the *end-vertices* of P: for instance, ξ_1 and ξ_7 are the end-vertices of the path in Fig. 5.

LEMMA 1: *Every tortuous graph has a hypertortuous spanned supergraph.*

Proof: Let G be a tortuous graph. Then G has at least one tortuous spanned supergraph, viz., G itself, and moreover any tortuous spanned supergraph H of G has the property that $|E(G)|$ $\leqslant |E(H)| \leqslant \binom{|V(G)|}{2}$ because $\binom{|V(G)|}{2}$, being the number of two-element subsets of $V(G)$, is the largest possible number of edges of a graph with the same vertices as G. Hence, amongst the integers $|E(G)|, |E(G)| + 1, |E(G)| + 2, \ldots, \binom{|V(G)|}{2}$, there must be a largest one (t, say) which is the number of edges of some tortuous spanned supergraph of G. Let H be a tortuous spanned supergraph of G with t edges. If any one-edge extension H' of H were tortuous, then H' would be a tortuous spanned supergraph of G with $t + 1$ edges, which would contradict the definition of t: hence H has no tortuous one-edge extension and so is hypertortuous. This proves that G has a hypertortuous spanned supergraph.

LEMMA 2. *If α, β are the end-vertices of a Hamiltonian path P of a graph G and $v(\alpha) + v(\beta) \geqslant |V(G)| \geqslant 3$, then G has a Hamiltonian circuit.*

Proof: Suppose that $\xi_1, \xi_2, \ldots, \xi_n$ are the vertices of G in the order in which they are encountered as we proceed along P from α to β, so that $\xi_1 = \alpha$, $\xi_n = \beta$ and, for each positive integer i less than n, an edge (λ_i, say) of P joins ξ_i to ξ_{i+1}. Since $V(G) = \{\xi_1, \xi_2, \ldots, \xi_n\}$ it follows that ξ_1 must be adjacent in G to $v(\xi_1)$ of the vertices $\xi_2, \ldots, \xi_n$ and hence the set (A, say) of those elements i of the set $\{1, 2, \ldots, n - 1\}$ for which ξ_1 adj ξ_{i+1} has cardinality $v(\xi_1)$. [Of course, ξ_1 is adjacent in G to ξ_2 since λ_1 joins these vertices; and so 1 is one of the numbers in the set A.] Furthermore, ξ_n must be adjacent in G to $v(\xi_n)$ of the vertices $\xi_1, \xi_2, \ldots, \xi_{n-1}$ and so the set (B, say) of those elements i of the set $\{1, 2, \ldots, n - 1\}$ for which ξ_n adj ξ_i has cardinality $v(\xi_n)$. Hence

$$|A| + |B| = v(\xi_1) + v(\xi_n) = v(\alpha) + v(\beta);$$

and by hypothesis

$$v(\alpha) + v(\beta) \geqslant |V(G)| = |\{\xi_1, \ldots, \xi_n\}| = n,$$

so that A and B are subsets of $\{1, 2, \ldots, n - 1\}$ such that $|A| + |B| \geqslant n$. It follows that A and B must have at least one element in common. Let I be a common element of A and B. Then ξ_1 adj ξ_{I+1} since $I \in A$ and ξ_n adj ξ_I since $I \in B$. Therefore there is a $\xi_1\xi_{I+1}$-edge μ and a $\xi_n\xi_I$-edge ν in G. It is now easily seen that $P - \lambda_I + \mu + \nu$, i.e., the subgraph of G obtained from P by removing the edge λ_I and adding the edges μ and ν, is a Hamiltonian circuit of G (cf. Fig. 6).

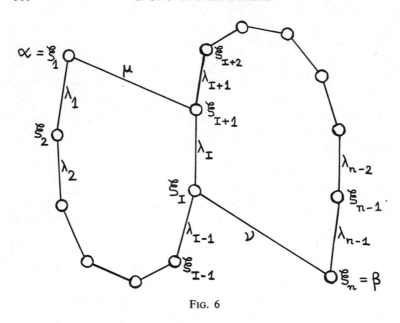

FIG. 6

The following slightly curious definition turns out to be very useful for our discussion: G will be said to be *stout* if α adj β for every two distinct vertices α, β of G such that $v(\alpha) + v(\beta) \geqslant |V(G)|$. In words, a stout graph is one in which every two distinct vertices, whose valencies add up to at least the number of vertices of the graph, are adjacent.

LEMMA 3: *Every hypertortuous graph is stout.*

Proof: Let G be hypertortuous and α, β be distinct vertices of G such that

$$v(\alpha) + v(\beta) \geqslant |V(G)|. \tag{1}$$

Then we must prove that α adj β, and we shall prove this by contradiction.

Suppose, therefore, that α is not adjacent to β. Then adding an edge λ joining α to β will transform G into a one-edge extension $G + \lambda$ of G, and, since G is hypertortuous, $G + \lambda$ must be round. Consequently $G + \lambda$ has a Hamiltonian circuit C, say. If $\lambda \notin E(C)$, then C is a Hamiltonian circuit of G. If $\lambda \in E(C)$ then $C - \lambda$ is a Hamiltonian path of G with end-vertices α, β, and the existence of this Hamiltonian path and (1) together imply, by Lemma 2, that G has a Hamiltonian circuit. Thus, whether λ belongs to $E(C)$ or not, we find that G has a Hamiltonian circuit, contradicting the hypothesis that G is hypertortuous. This contradiction shows that α must be adjacent to β, and the lemma is proved.

As we mentioned before, the first theorem on the lines of saying that "a graph has a Hamiltonian circuit if the valencies of its vertices are large enough" was given by Dirac [7]. We now state and prove Dirac's theorem.

THEOREM 1: *If* $|V(G)| = n \geqslant 3$ *and every vertex of* G *has valency* $\geqslant \frac{1}{2}n$, *then* G *has a Hamiltonian circuit.*

Proof: Let G be a graph such that $|V(G)| = n \geqslant 3$ and every vertex of G has valency $\geqslant \frac{1}{2}n$.

Suppose that G is tortuous. Then, by Lemma 1, G has a hypertortuous spanned supergraph (H, say). Then $V(G) = V(H)$ since H is a spanned supergraph of G, so that $|V(H)| = n$. Moreover, for every $\xi \in V(H)$, we have $v_H(\xi) \geqslant v_G(\xi) \geqslant \frac{1}{2}n$ and consequently, for every pair ξ, η of distinct vertices of H, we have $v_H(\xi) + v_H(\eta) \geqslant n = |V(H)|$. But H is by Lemma 3 stout. Hence every two distinct vertices of H are adjacent in H. From this it obviously follows that H has a Hamiltonian circuit and so is not tortuous. On the other hand, H is supposed to be hypertortuous, and therefore H is tortuous.

Thus the supposition that G is tortuous leads to a contradiction, whence we can conclude that G has a Hamiltonian circuit.

In Figures 7 and 8, we have shown two graphs with 11 vertices and no Hamiltonian circuits, in which every vertex has valency

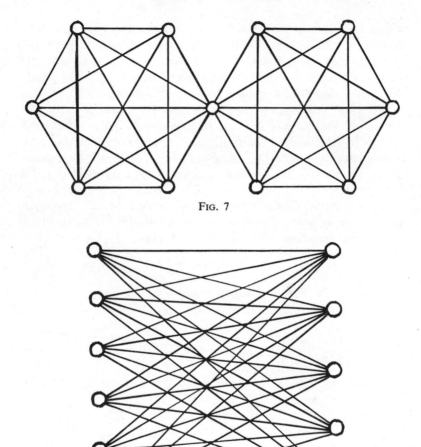

FIG. 7

FIG. 8

$\geqslant 5$. These and similar counterexamples show that, in Theorem 1, $\frac{1}{2}n$ cannot be reduced even as far as $(n-1)/2$. Nevertheless, it is possible to improve Theorem 1 in the direction of allowing *some but not all* of the vertices to have valencies somewhat less than $\frac{1}{2}n$

whilst still being able to conclude that the graph has a Hamiltonian circuit. To this end, we define the *valency sequence* $vs(G)$ of a graph G to be the sequence of numbers obtained by listing the valencies of the vertices of G in nondecreasing order: for example, the graph of Figure 9 has valency sequence 1, 1, 2, 2, 4 and that of Figure 7 has valency sequence 5, 5, 5, 5, 5, 5, 5, 5, 5, 5, 10. A nondecreasing sequence of numbers $a_1, \ldots, a_n$ is called *graphic* if it is the valency sequence of some graph: for example, the reader will readily see that the sequence 0, 0, 0, 0, 0, 0, 3, 3 is not graphic. [Remember that only *simple* graphs count as "graphs" for the purposes of our present discussion.] In fact, according to a theorem of Erdös and Gallai (see [13], Theorem 6.2), a nondecreasing sequence $a_1, \ldots, a_n$ of nonnegative integers is graphic iff $a_1 + \cdots + a_n$ is even and

$$\sum_{i=n-r+1}^{n} a_i \leqslant r(r-1) + \sum_{i=1}^{n-r} \min(a_i, r) \qquad (r = 1, 2, \ldots, n).$$

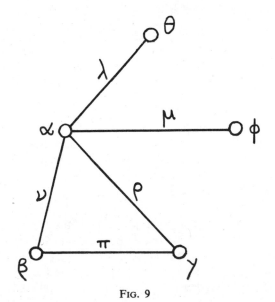

FIG. 9

Our next theorem, a considerable strengthening of Theorem 1 due to V. Chvátal, will concern graphs whose valency sequences satisfy a certain condition: we shall say that a valency sequence $a_1, a_2, \ldots, a_n$ satisfies *Chvátal's condition* if

$$\left. \begin{array}{l} \text{for each positive integer } i \text{ less than } n/2, \text{ at least} \\ \text{one of the inequalities } a_i \geqslant i + 1, a_{n-i} \geqslant n - i \text{ is true.} \end{array} \right\} \quad (2)$$

To illustrate this condition, let us suppose for example that $n = 31$, so that the positive integers less than $n/2$ are $1, 2, \ldots, 15$ and (2) asserts that, for each of these values of i, one or both of the inequalities $a_i \geqslant i + 1$, $a_{31-i} \geqslant 31 - i$ is true. So, if, for instance,

$$\left. \begin{array}{l} a_1 = a_2 = 3, a_3 = a_4 = a_5 = 4, a_6 = a_7 = 5, a_8 = 9, \\ a_9 = 10, \ a_{10} = 11, a_{11} = a_{12} = 12, a_{13} = 14, \\ a_{14} = a_{15} = a_{16} = a_{17} = a_{18} = 16, a_{19} = a_{20} = 20, \\ a_{21} = a_{22} = a_{23} = 22, a_{24} = a_{25} = 25, a_{26} = 26, \\ a_{27} = a_{28} = a_{29} = a_{30} = 27, a_{31} = 28, \end{array} \right\} \quad (3)$$

then the inequality $a_i \geqslant i + 1$ is satisfied for $i = 1, 2, 3, 8, 13, 14$, the inequality $a_{31-i} \geqslant 31 - i$ is satisfied for $i = 4, 5, 6, 7, 12$ and *both* of the inequalities $a_i \geqslant i + 1$, $a_{31-i} \geqslant 31 - i$ are satisfied for $i = 9, 10, 11, 15$. Thus the sequence $a_1, \ldots, a_{31}$ in (3) satisfies Chvátal's condition. If we took $n = 30$, then the positive integers less than $n/2$ would be $1, 2, \ldots, 14$ and so a sequence $a_1, \ldots, a_{30}$ would satisfy Chvátal's condition iff, for each of the values $1, 2, \ldots, 14$ of i, at least one of the inequalities $a_i \geqslant i + 1$, $a_{30-i} \geqslant 30 - i$ is true.

We notice also that if a graph G satisfies the hypotheses of Theorem 1, then its valency sequence satisfies Chvátal's condition. For the hypothesis that the valencies of the vertices of G are $\geqslant \frac{1}{2}n$ really means (since valencies are integers) that they are $\geqslant \frac{1}{2}n$ if n is even and $\geqslant \frac{1}{2}(n + 1)$ if n is odd, i.e., that $a_i \geqslant \frac{1}{2}n$

$(i = 1, 2, \ldots, n)$ if n is even and $a_i \geqslant \frac{1}{2}(n + 1)$ $(i = 1, 2, \ldots, n)$ if n is odd, where $a_1, \ldots, a_n$ is the valency sequence of G. This implies that $a_i \geqslant \frac{1}{2}n \geqslant i + 1$ for $i = 1, 2, \ldots, (n/2) - 1$ if n is even and $a_i \geqslant \frac{1}{2}(n + 1) \geqslant i + 1$ for $i = 1, 2, \ldots, (n - 1)/2$ if n is odd, so that, in both cases, $a_i \geqslant i + 1$ for every positive integer i less than $n/2$ and thus the valency sequence $a_1, \ldots, a_n$ satisfies (2).

Our next lemma records, for convenient reference, two very simple consequences of Chvátal's condition.

LEMMA 4: *Let* $a_1, \ldots, a_n$ *be the valency sequence of a graph. Suppose that* $n \geqslant 3$ *and* $a_1, \ldots, a_n$ *satisfies* (2). *Then*

(i) $a_i \geqslant 2$ *for* $i = 1, \ldots, n$;

(ii) $a_j \geqslant n/2$ *for every integer* j *such that* $n/2 < j \leqslant n$.

Proof: From the definition of what we mean by saying that $a_1, \ldots, a_n$ is the valency sequence of a graph (G, say), it follows that this sequence is nondecreasing, i.e.,

$$a_1 \leqslant a_2 \leqslant \cdots \leqslant a_n, \tag{4}$$

and that there is an ordering $\xi_1, \ldots, \xi_n$ of $V(G)$ such that

$$v(\xi_i) = a_i \qquad \text{for } i = 1, \ldots, n. \tag{5}$$

To prove (i), observe that, since (2) holds for $i = 1$, either $a_1 \geqslant 2$ or $a_{n-1} \geqslant n - 1$. If $a_1 \geqslant 2$, then (4) implies that all the a_i are $\geqslant 2$, i.e., that (i) holds: so we may assume that $a_{n-1} \geqslant n - 1$. Then, in view of (4), a_{n-1} and a_n are both $\geqslant n - 1$, i.e., (by (5)) $v(\xi_{n-1})$ and $v(\xi_n)$ are $\geqslant n - 1$, and this can only be achieved if ξ_{n-1} is joined by edges to *all* of the other $n - 1$ vertices $\xi_1, \xi_2, \ldots, \xi_{n-3}, \xi_{n-2}, \xi_n$ and ξ_n is joined to all of $\xi_1, \xi_2, \ldots, \xi_{n-2}, \xi_{n-1}$. Hence each of $\xi_1, \ldots, \xi_{n-2}$ must be joined by edges to both ξ_{n-1} and ξ_n, so that the valencies $a_1, \ldots, a_{n-2}$ of $\xi_1, \ldots, \xi_{n-2}$ must be $\geqslant 2$; and a_{n-1}, a_n are $\geqslant 2$ since they are $\geqslant n - 1$. This establishes (i).

To prove (ii), suppose that j is an integer such that $n/2 < j$

$\leqslant n$. Suppose, first, that n is even. Then $(n/2) - 1$ is a positive integer less than $n/2$ and so (2) holds for $i = (n/2) - 1$, i.e.,

$$\text{either} \quad a_{(n/2)-1} \geqslant \frac{n}{2} \quad \text{or} \quad a_{(n/2)+1} \geqslant \frac{n}{2} + 1. \tag{6}$$

But, since j is an integer greater than $n/2$ and n is even, it follows that $j \geqslant (n/2) + 1$ and consequently, by (4),

$$a_j \geqslant a_{(n/2)+1} \geqslant a_{(n/2)-1}. \tag{7}$$

From (6) and (7), it is clear that $a_j \geqslant n/2$. Now suppose that n is odd. Then $(n - 1)/2$ is a positive integer less than $n/2$ and so (2) holds for $i = (n - 1)/2$, i.e.,

$$\text{either} \quad a_{(n-1)/2} \geqslant (n + 1)/2 \quad \text{or} \quad a_{(n+1)/2} \geqslant (n + 1)/2; \tag{8}$$

and, since $a_{(n+1)/2} \geqslant a_{(n-1)/2}$ by (4), the first alternative in (8) implies the second and so we have $a_{(n+1)/2} \geqslant (n + 1)/2$ in any case. But, since j is an integer greater than $n/2$, it follows that $j \geqslant (n + 1)/2$ and so, by (4), $a_j \geqslant a_{(n+1)/2} \geqslant (n + 1)/2 > n/2$. Thus, considering first the case in which n is even and then the case in which n is odd, we have shown that $a_j \geqslant n/2$ for each integer j such that $n/2 < j \leqslant n$.

DEFINITION: Let s denote a finite sequence of numbers $a_1, \ldots, a_n$ and t denote a finite sequence of numbers $b_1, \ldots, b_n$ with the same number of terms as s. Then the statement $s \leqslant t$ will mean that $a_i \leqslant b_i$ for $i = 1, \ldots, n$, i.e., each term of s is less than or equal to the corresponding term of t.

LEMMA 5: *If G is a spanning subgraph of H, then $vs(G) \leqslant vs(H)$.*

Proof: Let $\xi_1, \ldots, \xi_n$ be an ordering of $V(G)$ such that

$$v_G(\xi_1) \leqslant v_G(\xi_2) \leqslant \ldots \leqslant v_G(\xi_n) \tag{9}$$

and $\eta_1, \ldots, \eta_n$ be an ordering of $V(H)$ $(= V(G))$ such that

$$v_H(\eta_1) \leqslant v_H(\eta_2) \leqslant \ldots \leqslant v_H(\eta_n). \qquad (10)$$

Consider any $r \in \{1, \ldots, n\}$. Since $|V(G)| = n$ and the sum of the cardinalities of the subsets $\{\xi_r, \xi_{r+1}, \ldots, \xi_n\}$ and $\{\eta_1, \eta_2, \ldots, \eta_r\}$ of $V(G)$ is $n + 1$, these two subsets cannot be disjoint, and so $\xi_i = \eta_j$ for some $i \geqslant r$ and some $j \leqslant r$. Then since G is a subgraph of H, we have $v_G(\xi_i) \leqslant v_H(\xi_i) = v_H(\eta_j)$: combining this with the facts that $v_G(\xi_r) \leqslant v_G(\xi_i)$ by (9) and $v_H(\eta_j) \leqslant v_H(\eta_r)$ by (10), we conclude that $v_G(\xi_r) \leqslant v_H(\eta_r)$. This argument proves that $v_G(\xi_r) \leqslant v_H(\eta_r)$ for $r = 1, \ldots, n$, and since $v_G(\xi_r)$, $v_H(\eta_r)$ are the rth terms of $vs(G)$, $vs(H)$ respectively we conclude that $vs(G) \leqslant vs(H)$.

For example, if H is the graph of Fig. 10 and G is the spanning subgraph of H shown in Fig. 9, then $vs(G)$ is 1, 1, 2, 2, 4 and $vs(H)$ is 2, 2, 3, 3, 4 and the inequalities $1 \leqslant 2$, $1 \leqslant 2$, $2 \leqslant 3$, $2 \leqslant 3$, $4 \leqslant 4$ show that $vs(G) \leqslant vs(H)$ as predicted by Lemma 5.

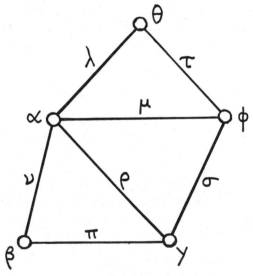

FIG. 10

We notice, however, that the sequences 1, 1, 2, 2, 4 and 2, 2, 3, 3, 4 must be derived from different orderings of the vertices: for example, the former sequence can be thought of as $v_G(\theta)$, $v_G(\phi)$, $v_G(\beta)$, $v_G(\gamma)$, $v_G(\alpha)$ whilst the latter is $v_H(\theta)$, $v_H(\beta)$, $v_H(\phi)$, $v_H(\gamma)$, $v_H(\alpha)$. Thus Lemma 5 is not just a trivial matter of observing that $v_G(\xi) \leqslant v_H(\xi)$ for each vertex ξ: the statement that $vs(G) \leqslant vs(H)$ may entail comparisons of valencies of certain vertices in G with valencies of *different* vertices in H.

LEMMA 6: *If G is stout and $|V(G)| \geqslant 3$ and $vs(G)$ satisfies Chvátal's condition, then every two distinct vertices of G are adjacent.*

Proof: The valency sequence of G can be written as $v(\xi_1)$, $v(\xi_2)$, ..., $v(\xi_n)$ where ξ_1, ..., ξ_n is an ordering of $V(G)$ such that

$$v(\xi_1) \leqslant v(\xi_2) \leqslant \ldots \leqslant v(\xi_n). \tag{11}$$

Since this valency sequence satisfies Chvátal's condition (2), we know that

$$\left. \begin{array}{l} \text{for each positive integer } i \text{ less than } n/2, \text{ at least one} \\ \text{of the inequalities } v(\xi_i) \geqslant i + 1, v(\xi_{n-i}) \geqslant n - i \text{ is true.} \end{array} \right\} \tag{12}$$

Let S_r denote the statement that every two distinct vertices in the set $\{\xi_r, \xi_{r+1}, \xi_{r+2}, \ldots, \xi_n\}$ are adjacent. Thus S_1 is the statement that *every* two distinct vertices of G are adjacent, which is what we wish to prove. However, by Lemma 4(ii), any two distinct vertices ξ_j ($j > n/2$) and ξ_k ($k > n/2$) have valency-sum $\geqslant n = |V(G)|$ and are consequently adjacent since G is stout. Therefore $S_{(n/2)+1}$ is true if n is even and $S_{(n+1)/2}$ is true if n is odd. So the desired conclusion S_1 can be established if we show that

$$S_{(n/2)+1} \Rightarrow S_{n/2} \Rightarrow S_{(n/2)-1} \Rightarrow \cdots \Rightarrow S_2 \Rightarrow S_1$$

if n is even and that

$$S_{(n+1)/2} \Rightarrow S_{(n-1)/2} \Rightarrow S_{(n-3)/2} \Rightarrow \cdots \Rightarrow S_2 \Rightarrow S_1$$

if n is odd, i.e., we need to prove that $S_{m+1} \Rightarrow S_m$ whenever m is a positive integer $\leqslant n/2$.

Therefore, suppose that m is a positive integer $\leqslant n/2$ and that S_{m+1} is true. Since S_{m+1} is true, the $n - m$ vertices ξ_{m+1}, $\xi_{m+2}, \ldots, \xi_n$ are pairwise adjacent, so that each of them, being joined by edges to the $n - m - 1$ others, has valency $\geqslant n - m - 1$. Hence, if $v(\xi_m) \geqslant m + 1$, then the sums

$$v(\xi_m) + v(\xi_{m+1}), v(\xi_m) + v(\xi_{m+2}), \ldots, v(\xi_m) + v(\xi_n) \quad (13)$$

will be $\geqslant n$ and so ξ_m will be adjacent to each of ξ_{m+1}, $\xi_{m+2}, \ldots, \xi_n$, and this fact combined with the pairwise adjacency of $\xi_{m+1}, \xi_{m+2}, \ldots, \xi_n$ will imply that S_m is true, as required. Also, if $v(\xi_m) \geqslant n/2$ then by (11) the valencies $v(\xi_m), v(\xi_{m+1}), \ldots, v(\xi_n)$ will all be $\geqslant n/2$ and so the sums (13) will all be $\geqslant n$, from which S_m will follow as before. So it suffices to prove that either $v(\xi_m) \geqslant m + 1$ or $v(\xi_m) \geqslant n/2$. [Of course the latter of these alternatives is weaker than the former only when $m = n/2$.] We prove this by contradiction: suppose that $v(\xi_m)$ is an integer p less than $m + 1$ and less than $n/2$. This integer $p = v(\xi_m)$ is by Lemma 4(i) a *positive* integer less than $n/2$, so that by (12) either $v(\xi_p) \geqslant p + 1$ or $v(\xi_{n-p}) \geqslant n - p$. But since $p < m + 1$, it follows that $p \leqslant m$ and hence, by (11), $v(\xi_p) \leqslant v(\xi_m) = p$, which rules out the alternative $v(\xi_p) \geqslant p + 1$. So we must have $v(\xi_{n-p}) \geqslant n - p$. From this inequality and (11), it follows that the vertices ξ_{n-p}, $\xi_{n-p+1}, \ldots, \xi_n$ all have valencies $\geqslant n - p$, and so, since $v(\xi_m) = p$, the sums

$$v(\xi_m) + v(\xi_{n-p}), v(\xi_m) + v(\xi_{n-p+1}), \ldots, v(\xi_m) + v(\xi_n)$$

are all $\geqslant n$. Moreover, ξ_m is not one of $\xi_{n-p}, \xi_{n-p+1}, \ldots, \xi_n$ since $m \leqslant n/2$ and $p < n/2$. These facts imply, since G is stout, that ξ_m is joined by $p + 1$ edges to the $p + 1$ vertices $\xi_{n-p}, \xi_{n-p+1}, \ldots, \xi_n$, contradicting the fact that $v(\xi_m) = p$. Thus the supposition that $v(\xi_m)$ is less than both $m + 1$ and $n/2$ leads to a contradiction, i.e., we must have $v(\xi_m) \geqslant m + 1$ or $v(\xi_m) \geqslant n/2$, and we have shown that either of these inequalities implies the truth of S_m. So the implication $S_{m+1} \Rightarrow S_m$ has been proved for all positive integers $m \leqslant n/2$, thus completing the proof of Lemma 6.

THEOREM 2: (Chvátal's Theorem). *If $|V(G)| \geqslant 3$ and $vs(G)$ satisfies Chvátal's condition, then G has a Hamiltonian circuit.*

Proof: Let G be a graph such that $|V(G)| = n \geqslant 3$ and $vs(G)$ is a sequence $a_1, \ldots, a_n$ satisfying Chvátal's condition (2).

Suppose that G is tortuous. Then, by Lemma 1, G has a hypertortuous spanned supergraph (H, say). By Lemma 5, $vs(G) \leqslant vs(H)$, i.e., $vs(H)$ is a sequence $b_1, b_2, \ldots, b_n$ such that

$$a_1 \leqslant b_1, a_2 \leqslant b_2, \ldots, a_n \leqslant b_n. \tag{14}$$

From (14) and the condition (2) satisfied by $a_1, \ldots, a_n$, it follows that for each positive integer i less than $n/2$, we have either $b_i \geqslant a_i \geqslant i + 1$ or $b_{n-i} \geqslant a_{n-i} \geqslant n - i$ (or both), so that the valency sequence $b_1, \ldots, b_n$ of H also satisfies Chvátal's condition. Moreover, H is stout by Lemma 3. Since H is stout and $|V(H)| = |V(G)| \geqslant 3$ and $vs(H)$ satisfies Chvátal's condition, it follows by Lemma 6 that every two distinct vertices of H are adjacent in H. This leads to a contradiction exactly as in the proof of Theorem 1, showing that G cannot have been tortuous and consequently must have a Hamiltonian circuit.

As a corollary to Theorem 2, we deduce Pósa's Theorem. This procedure actually reverses the historical order of events: Pósa's Theorem was discovered some eight years before Theorem 2 and was almost certainly a major contributory cause of various subsequent developments including an improved version of Pósa's Theorem due to J. A. Bondy and then the still stronger Theorem 2. To give a convenient statement of Pósa's Theorem, we define P_n to be a sequence of n numbers defined as follows:

(i) if n is even, P_n is the sequence

$$2, 3, 4, 5, \ldots, \frac{n}{2} - 2, \frac{n}{2} - 1, \frac{n}{2}, \frac{n}{2}, \frac{n}{2}, \frac{n}{2}, \ldots, \frac{n}{2}, \frac{n}{2};$$

(ii) if n is odd, P_n is the sequence

$$2, 3, 4, 5, \ldots, \frac{n-5}{2}, \frac{n-3}{2}, \frac{n-1}{2}, \frac{n-1}{2},$$

$$\frac{n+1}{2}, \frac{n+1}{2}, \frac{n+1}{2}, \ldots, \frac{n+1}{2}, \frac{n+1}{2}.$$

It is to be understood that the number of appearances of $n/2$ at the end of the sequence in (i) and the number of appearances of $(n + 1)/2$ in (ii) is such as to make the total number of terms equal to n, i.e., $(n/2) + 2$ appearances and $(n + 1)/2$ appearances respectively. Since this definition is not too clear for small values of n, we state that P_3 is the sequence 1, 2, 2 and P_4 is 2, 2, 2, 2 and P_5 is 2, 2, 3, 3, 3 and P_6 is 2, 3, 3, 3, 3, 3 and P_7 is 2, 3, 3, 4, 4, 4, 4.

COROLLARY 2A: (Pósa's Theorem). *If* $|V(G)| = n \geqslant 3$ *and* $vs(G) \geqslant P_n$, *then* G *has a Hamiltonian circuit.*

Proof: Let $vs(G)$ be $a_1, \ldots, a_n$. If n is even, the ith term of P_n is $i + 1$ for every positive integer i less than $n/2$ and consequently the hypothesis $vs(G) \geqslant P_n$ implies that $a_i \geqslant i + 1$ for every positive integer i less than $n/2$: therefore the sequence $a_1, \ldots, a_n$ satisfies (2). If n is odd, the ith term of P_n is $i + 1$ for every positive integer i less than $(n - 1)/2$ and, for $i = (n - 1)/2$, the $(n - i)$th term of P_n, i.e., its $(n + 1)/2$th term, is $(n + 1)/2$, i.e., $n - i$. Thus, in this case, the hypothesis $vs(G) \geqslant P_n$ implies that $a_i \geqslant i + 1$ for every positive integer $i < (n - 1)/2$ and $a_{n-i} \geqslant n - i$ for $i = (n - 1)/2$, so that once more the sequence $a_1, \ldots, a_n$ satisfies (2). Thus, whether n be even or odd, the valency sequence $a_1, \ldots, a_n$ of G satisfies Chvátal's condition (2) and consequently by Theorem 2, G has a Hamiltonian circuit. If $|V(G)| = n \geqslant 3$ and every vertex of G has valency $\geqslant \frac{1}{2}n$, then in fact, because the valencies of the vertices of G are integers, they are $\geqslant \frac{1}{2}n$ if n is even and $\geqslant \frac{1}{2}(n + 1)$ if n is odd, and hence $vs(G) \geqslant P_n$. This makes it clear that Dirac's Theorem is contained in Pósa's Theorem, which in turn is shown by the proof of Corollary 2A to be contained in Chvátal's Theorem. Obviously Pósa's Theorem is in fact stronger than Dirac's Theorem, and valency sequences such as (3), which satisfy Chvátal's condition but not the hypotheses of Pósa's Theorem, show that Chvátal's Theorem is strictly stronger than Pósa's.

3. DIRECTED GRAPHS

In this section, we consider directed graphs, commonly referred to as *digraphs*, and examine briefly the extent to which they give rise to results and problems analogous to those of Section 2. In a directed graph, each edge is associated with an *ordered* pair of vertices, which it is said to *join*, the first member of the ordered pair being called the *tail* of the edge and the second member of the ordered pair being called its *head*. We denote the tail and head of an edge λ by λt, λh respectively. In diagrams representing digraphs (e.g., Fig. 11) we place an arrow on each edge, pointing in the direction from the tail to the head of the edge. The *outvalency* $v_{\text{out}}(\xi)$ of a vertex ξ of a digraph is the number of edges with tail ξ and its *invalency* $v_{\text{in}}(\xi)$ is the number of edges with head ξ. The *valency* $v(\xi)$ of ξ is the total number of edges incident with ξ, i.e., $v(\xi) = v_{\text{in}}(\xi) + v_{\text{out}}(\xi)$. The *outvalency sequence* $ovs(D)$ of the digraph D is the sequence of numbers obtained by listing the outvalencies of the vertices of D in nondecreasing order, and the *invalency sequence* $ivs(D)$ of D is obtained by listing the invalencies of its vertices in nondecreasing order: for example, the digraph of Fig. 12 has outvalency sequence 2, 2, 2, 4, 4, 5 and invalency sequence 1, 2, 4, 4, 4, 4. We call a digraph D *disimple* if (i) for every edge λ of D, $\lambda t \neq \lambda h$ and (ii) D does not possess two

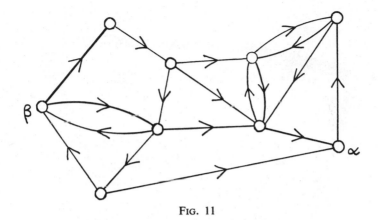

FIG. 11

edges λ, μ such that $\lambda t = \mu t$ and $\lambda h = \mu h$. However, two distinct vertices ξ, η of a disimple digraph can be joined by 0, 1 or 2 edges, since (ii) does not preclude the existence of two edges λ, μ such that $\lambda t = \mu h = \xi$, $\lambda h = \mu t = \eta$. All digraphs considered in Section 3 will be understood to be finite (in the sense that they have only finitely many vertices and edges) and disimple. The letter D will always denote a digraph.

In general, terminology and notation relating to digraphs should, in the absence of a special definition, be understood to be used in the same sense as when it is applied to graphs: for instance, $V(D)$ denotes the set of vertices of a digraph D.

A non-empty digraph P is a *dipath* (or *directed path*) if there is an ordering $\xi_1, \xi_2, \ldots, \xi_n$ of $V(P)$ and an ordering $\lambda_1, \lambda_2, \ldots, \lambda_{n-1}$ of $E(P)$ such that $\lambda_i t = \xi_i$ and $\lambda_i h = \xi_{i+1}$ for $i = 1, 2, \ldots, n - 1$: we call ξ_1 the *initial vertex* of P and ξ_n its *terminal vertex*. Fig. 13 illustrates this definition for $n = 6$. [A digraph with just one vertex ξ and no edge counts as a dipath whose initial and terminal vertex are both ξ.] A non-empty digraph C is a *dicircuit* (or *directed circuit*) if there is an ordering $\xi_1, \xi_2, \ldots, \xi_n$ of $V(C)$ and an ordering $\lambda_1, \lambda_2, \ldots, \lambda_n$ of $E(C)$ such that $\lambda_i t = \xi_i$ and $\lambda_i h = \xi_{i+1}$ for $i = 1, 2, \ldots, n - 1$ and $\lambda_n t = \xi_n$ and $\lambda_n h = \xi_1$. Fig. 14 depicts dicircuits with 2, 3, 4 and 5

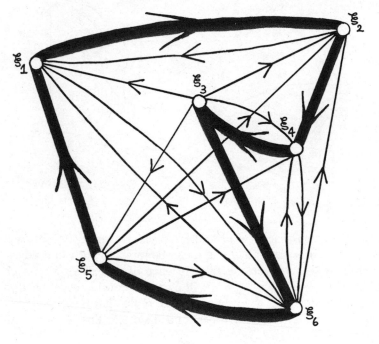

FIG. 12

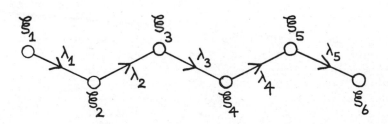

FIG. 13

322

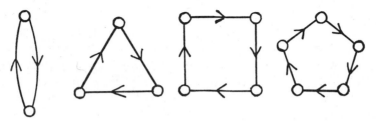

vertices. (Since only disimple digraphs are allowed in our discussion, a dicircuit with 1 vertex is impossible.) A *Hamiltonian dicircuit* of a digraph D is a dicircuit, contained in D, which includes all the vertices of D: for example, if D is the digraph depicted in Fig. 12, then D has a Hamiltonian dicircuit made up of all the vertices of D and those edges which are shown in the diagram by thick lines.

By analogy with our discussion of Hamiltonian circuits in graphs, one can ask questions about which digraphs have Hamiltonian dicircuits. In particular, the line of thinking pursued in Section 2 raises the analogous question of what we can learn about the existence of Hamiltonian dicircuits in a digraph from information about the invalencies, outvalencies and valencies of its vertices; and it transpires that the following analogue of Dirac's theorem is true.

THEOREM 3: *If* $|V(D)| = n > 2$ *and every vertex of D has invalency* $> \frac{1}{2}n$ *and outvalency* $> \frac{1}{2}n$, *then D has a Hamiltonian dicircuit.*

In fact, there is a stronger theorem, for whose statement we need first the definition of a strongly connected or (as the present author prefers to call it) diconnected digraph. A digraph D is *diconnected* if, for every ordered pair (ξ, η) of vertices of D, there exists a dipath in D with initial vertex ξ and terminal vertex η. For instance, the digraph in Fig. 11 is not diconnected since it contains no dipath with initial vertex α and terminal vertex β. On the other hand, if a digraph D satisfies the hypotheses of Theorem 3, then it is diconnected. This can be established either by an appeal to Theorem 3 or, much more simply, by the following direct argument. Suppose that $|V(D)| = n \geqslant 2$ and every vertex of D has invalency $\geqslant \frac{1}{2}n$ and outvalency $\geqslant \frac{1}{2}n$. Let (ξ, η) be an ordered pair of vertices of D. We wish to prove that there is a dipath with initial vertex ξ and terminal vertex η. If $\xi = \eta$, then the dipath P such that $V(P) = \{\xi\} = \{\eta\}$ and $E(P) = \varnothing$ has initial vertex ξ and terminal vertex η. If some edge λ of D has tail ξ and head η, then ξ, λ and η make up a dipath with initial vertex ξ and terminal vertex η. Now consider the remaining case, in which $\xi \neq \eta$ and D has no edge with tail ξ and head η. Then, if $\lambda_1, \ldots, \lambda_r$ are the edges of D with tail ξ and $\mu_1, \ldots, \mu_s$ are the edges of D with head η, the r distinct vertices $\lambda_1 h, \ldots, \lambda_r h$ and the s distinct vertices $\mu_1 t, \ldots, \mu_s t$ all belong to $V(G) \setminus \{\xi, \eta\}$. But $r = v_{\text{out}}(\xi) \geqslant n/2$ since r is the number of edges with tail ξ, and $s = v_{\text{in}}(\eta) \geqslant n/2$, and so $r + s \geqslant n = |V(G)| > |V(G) \setminus \{\xi, \eta\}|$. Hence one of $\lambda_1 h, \ldots, \lambda_r h$ must be the same as one of $\mu_1 t, \ldots, \mu_s t$, and so $\lambda_i h = \mu_j t$ for some i and j, with the result that ξ, λ_i, $\lambda_i h$, μ_j and η make up a dipath with initial vertex ξ and terminal vertex η.

Intuitively, it is natural to think of edges of a digraph as one-way streets; and from this point of view a diconnected digraph is one in which we can drive from any vertex to any other without violating traffic regulations.

The following ingenious result was proved by Ghouila-Houri [12].

THEOREM 4: *If $|V(D)| = n \geqslant 2$ and every vertex of D has valency $\geqslant n$ and D is diconnected, then D has a Hamiltonian dicircuit.*

Since the valency of a vertex of a digraph is the sum of its invalency and outvalency, it follows that, if D satisfies the hypotheses of Theorem 3, then the valencies of its vertices are $\geqslant n$. We have also observed that any digraph which satisfies the hypotheses of Theorem 3 is diconnected. Thus digraphs which

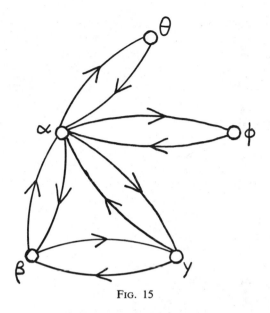

FIG. 15

satisfy the hypotheses of Theorem 3 automatically satisfy those of
Theorem 4, and so Theorem 4 contains Theorem 3. But Theorem 4
is considerably stronger, since many digraphs, such as that of Fig.
12, satisfy the hypotheses of Theorem 4 but not those of Theorem
3.

Not only does Theorem 4 contain Theorem 3, but also Theorem
3 in turn contains Theorem 1 in a certain sense. To explain this
remark, we define the notion of a digraph *equivalent* to a graph G.
This is a digraph D such that (i) $V(D) = V(G)$, (ii) if two vertices
are non-adjacent in G, then they are not joined by any edge of D,
and (iii) if two vertices ξ, η are adjacent in G, then they are joined
in D by *two* edges, of which one has tail ξ and head η whilst the
other has tail η and head ξ. For example, the digraph of Fig. 15 is
equivalent to the graph of Fig. 9. Intuitively, if one thinks of edges
of a graph as two-way streets and edges of a digraph as one-way
streets, this notion of "equivalence" is a natural one. To join two
vertices by two one-way streets permitting travel in opposite
directions is, for travelling purposes, "equivalent" to joining them

by a two-way street, and so a digraph equivalent to a graph G can be thought of as a network of one-way streets capable of replacing the network G of two-way streets.

If D is equivalent to a graph G with at least 3 vertices, then it is easily seen that G has a Hamiltonian circuit if and only if D has a Hamiltonian dicircuit, and also that the valency in G of any vertex is equal to both the invalency and the outvalency of that vertex in D. Thus Theorem 3 contains Theorem 1 in the sense that, if G is a graph satisfying the hypotheses of Theorem 1, we have only to construct a digraph D equivalent to G and observe that D then satisfies the hypotheses of Theorem 3 and so has, by Theorem 3, a Hamiltonian dicircuit, implying that G has a Hamiltonian circuit. In other words, Theorem 1 can be considered as the special case of Theorem 3 in which the digraph concerned is of such a kind that it is equivalent to some graph. In fact a digraph is called *symmetric* if it is equivalent to some graph; or, to express this definition in another way, a *symmetric* digraph is one in which every two vertices are joined *either* by two edges such that the tail of each of these edges is the head of the other *or* by no edge at all (Fig. 15).

Theorem 3 might be thought of as the analogue of Dirac's Theorem (Theorem 1) for digraphs, and I was motivated by this to conjecture in [14] that the following digraph analogue of Pósa's Theorem (Corollary 2A) might be true.

CONJECTURE 1: *If a digraph D has $n(\geqslant 3)$ vertices and $ivs(D)$ $\geqslant P_n$ and $ovs(D) \geqslant P_n$, then D has a Hamiltonian dicircuit.*

If Conjecture 1 is true, Pósa's Theorem could be deduced from it in the same way that Dirac's Theorem is deducible from Theorem 3; but until now nobody has either proved or disproved Conjecture 1, and the problem may be a very difficult one. It is possible to propose also quite a number of variants of Conjecture 1; *inter alia*, one might generalize it to some stronger conjecture in something like the way in which Theorem 4 generalizes Theorem 3. Again, one might propose a conjecture about digraphs which generalizes Chvátal's Theorem, possibly on some such lines as the following:

CONJECTURE 2: *If a diconnected digraph D has n(≥ 3) vertices and invalency sequence $a_1, \ldots, a_n$ and outvalency sequence $b_1, \ldots, b_n$ and if*

(i) *for every positive integer $i < n/2$, at least one of the inequalities $a_i \geq i + 1$, $b_{n-i} \geq n - i$ is true, and*

(ii) *for every positive integer $i < n/2$, at least one of the inequalities $b_i \geq i + 1$, $a_{n-i} \geq n - i$ is true,*

then D has a Hamiltonian dicircuit.

To conclude this section, it may be worth remarking that "which graphs have Hamiltonian circuits?" and "which digraphs have Hamiltonian dicircuits?" are equivalent questions in the sense that a complete answer to either of them would provide a complete answer to the other. For, if we knew which digraphs have Hamiltonian dicircuits, then, to determine whether a graph G had a Hamiltonian circuit, we would only have to consider a digraph Δ_G equivalent to G (in the sense defined above) and observe that G has a Hamiltonian circuit if and only if Δ_G has a Hamiltonian dicircuit: so our knowledge of whether or not Δ_G has a Hamiltonian dicircuit would answer the question about G also. Conversely, if we knew which graphs had Hamiltonian circuits then, to determine whether a given digraph D had a Hamiltonian dicircuit, we could use a certain construction to form a graph Γ_D which has a Hamiltonian circuit if and only if D has a Hamiltonian dicircuit: so our knowledge of whether Γ_D has a Hamiltonian circuit would answer the question about D also. The method of constructing Γ_D from D is illustrated by Figs. 12 and 16. If D has vertices $\xi_1, \xi_2, \ldots, \xi_n$, then Γ_D has vertices

$$\theta_1, \theta_2, \ldots, \theta_n, \phi_1, \phi_2, \ldots, \phi_n, \psi_1, \psi_2, \ldots, \psi_n;$$

and the edges of Γ_D are inserted as follows: join θ_i to ϕ_i by an edge and ϕ_i to ψ_i by an edge for $i = 1, 2, \ldots, n$ and, in addition, whenever D has an edge with tail ξ_i and head ξ_j, join ψ_i to θ_j by an edge of Γ_D. For the example of this construction represented by our diagrams, Fig. 12 exhibits, by means of thick edges, a Hamil-

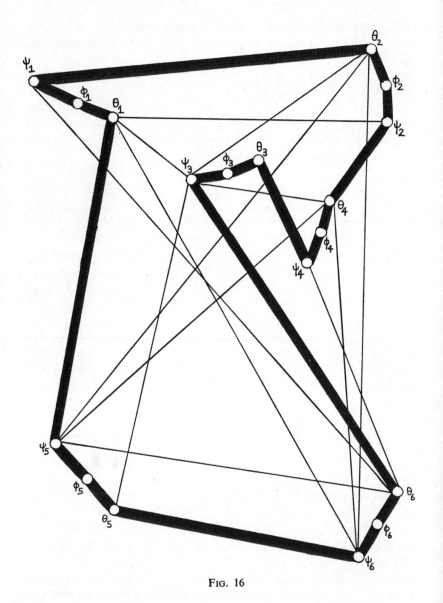

Fig. 16

tonian dicircuit in D and Fig. 16 exhibits, by means of thick edges, the "corresponding" Hamiltonian circuit in Γ_D. These diagrams should provide any interested reader with a sufficient hint for proving that, in general, Γ_D has a Hamiltonian circuit if and only if D has a Hamiltonian dicircuit. Unfortunately this construction does not permit us to deduce Conjecture 1 from the "corresponding" result (Corollary 2A) about graphs, because a digraph D satisfying the hypotheses of Conjecture 1 gives rise to a graph Γ_D which does not satisfy the hypotheses of Corollary 2A.

4. A WEAKER PROPERTY THAN HAVING A HAMILTONIAN CIRCUIT

The *length* $\ell(P)$ of a path P is the number of edges in P. A path whose end-vertices are ξ and η will be called a $\xi\eta$-*path*. The *distance* $d(\xi, \eta)$ between two distinct vertices ξ, η of a connected graph G is the minimum of the lengths of all $\xi\eta$-paths in G, i.e., intuitively, the minimum number of edges along which one would have to travel in order to get from ξ to η. For instance, in the graph of Fig. 2, $d(\alpha, \beta) = 1$, $d(\rho, \delta) = 2$, $d(\theta, \gamma) = 3$. To say that a graph G (with at least 3 vertices) has a Hamiltonian circuit is clearly equivalent to saying that its vertices can be arranged in a sequence $\xi_1, \xi_2, \ldots, \xi_n$ such that the distances

$$d(\xi_1, \xi_2), d(\xi_2, \xi_3), \ldots, d(\xi_{n-2}, \xi_{n-1}), d(\xi_{n-1}, \xi_n), d(\xi_n, \xi_1)$$

are all 1. For example, the existence of the Hamiltonian circuit indicated by the thick edges in the graph of Fig. 2 corresponds to the fact that the vertices of the graph can be arranged in the sequence $\alpha, \rho, \beta, \gamma, \sigma, \delta, \phi, \psi, \theta$ which has the property that the distances $d(\alpha, \rho)$, $d(\rho, \beta)$, $d(\beta, \gamma)$, $d(\gamma, \sigma)$, $d(\sigma, \delta)$, $d(\delta, \phi)$, $d(\phi, \psi)$, $d(\psi, \theta)$, $d(\theta, \alpha)$ are all 1. Thus we can identify a property of a graph which is in general weaker than having a Hamiltonian circuit by saying that a connected graph G is k-*round* (where k denotes a positive integer) if there exists an ordering $\xi_1, \xi_2, \ldots, \xi_n$

of $V(G)$ such that the distances

$$d(\xi_1, \xi_2), d(\xi_2, \xi_3), \ldots, d(\xi_{n-2}, \xi_{n-1}), d(\xi_{n-1}, \xi_n), d(\xi_n, \xi_1)$$

are all $\leqslant k$: then, provided that $|V(G)| \geqslant 3$, 1-roundness of G is equivalent to its having a Hamiltonian circuit and, for any $k > 1$, k-roundness is a weaker property. Thus, deciding which connected graphs are 1-round is equivalent to the very difficult problem of determining which of them have Hamiltonian circuits. Can we expect more luck in enquiring which connected graphs are 2-round, which ones are 3-round, etc.? It turns out (see Theorem 6 below) that *all* connected graphs are 3-round (and hence also k-round for all $k \geqslant 3$); but which of them are 2-round seems to be a non-trivial problem, which is not yet fully settled, although, as we shall indicate, very interesting progress has recently been made with it.

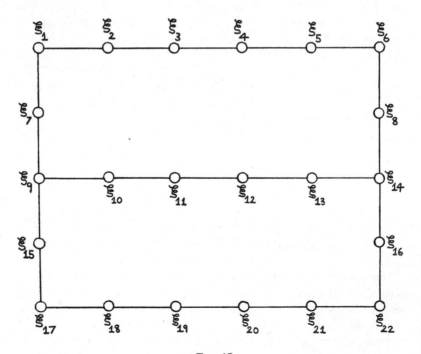

Fig. 17

A few definitions at this point will help us to proceed with clarity and precision. Throughout Section 4, the word *sequence* will mean "finite sequence." A sequence *on* a set S is a sequence all of whose terms belong to S: for example, the sequence 2, 3, 5, 3, 7 is a sequence on the set $\{1, 2, 3, 4, 5, 6, 7, 8\}$. Our earlier definitions of "$\xi\eta$-path" and of the "distance" $d(\xi, \eta)$ between vertices ξ and η will be extended to the case in which $\xi = \eta$ by defining a $\xi\xi$-*path* to be a path P such that $V(P) = \{\xi\}$ and $E(P) = \varnothing$ and defining the *distance* $d(\xi, \xi)$ between ξ and itself to be the length of this path, i.e., zero. Let G be a connected graph and k be a positive integer. Then a sequence $\xi_1, \xi_2, \ldots, \xi_n$ on $V(G)$ is k-*gradual* if the distances

$$d(\xi_1, \xi_2), d(\xi_2, \xi_3), \ldots, d(\xi_{n-2}, \xi_{n-1}), d(\xi_{n-1}, \xi_n)$$

are $\leqslant k$. If these distances and the distance $d(\xi_n, \xi_1)$ are all $\leqslant k$, we say that the sequence $\xi_1, \xi_2, \ldots, \xi_n$ is k-*cyclic*. [If $n = 1$, i.e., if ξ_1 is the sole term of the sequence "$\xi_1, \xi_2, \ldots, \xi_n$", then this sequence is considered to be both k-gradual and k-cyclic for all positive integers k.] Thus, in the graph of Fig. 17, the sequence

$$\xi_2, \xi_5, \xi_4, \xi_8, \xi_{16}, \xi_{20}, \xi_{17}, \xi_{10}$$

is 3-gradual. It is not 3-cyclic because $d(\xi_{10}, \xi_2) = 4$; but it is 4-cyclic. A connected graph G will be called k-*orderable* if $V(G)$ has a k-gradual ordering. We observe also that our definition of "k-round" amounts to saying that a connected graph G is k-round iff $V(G)$ has a k-cyclic ordering. The graph of Fig. 17 is 2-round, since, for example,

$$\left.\begin{array}{l} \xi_1, \xi_2, \xi_3, \xi_4, \xi_5, \xi_6, \xi_8, \xi_{14}, \xi_{13}, \xi_{12}, \xi_{11}, \xi_{10}, \xi_9, \\ \xi_{17}, \xi_{19}, \xi_{21}, \xi_{16}, \xi_{22}, \xi_{20}, \xi_{18}, \xi_{15}, \xi_7 \end{array}\right\} \quad (15)$$

is one possible 2-cyclic ordering of its set of vertices. The graph of Fig. 18 is 3-round, one possible 3-cyclic ordering of its set of vertices being

$$\alpha_1, \alpha_3, \alpha_5, \alpha_6, \alpha_4, \alpha_2, \beta_1, \beta_3, \beta_5, \beta_6, \beta_4, \beta_2, \omega, \gamma_2, \gamma_4, \gamma_6,$$
$$\gamma_5, \gamma_3, \gamma_1, \delta_1, \delta_3, \delta_5, \delta_6, \delta_4, \delta_2, \epsilon_1, \epsilon_3, \epsilon_5, \epsilon_6, \epsilon_4, \epsilon_2,$$

but the reader will probably fairly quickly convince himself by experiment that this graph is not 2-round, nor even 2-orderable. It will be convenient to make the convention that the empty graph (i.e., the graph which has no vertices and no edges) is considered to be k-round for every positive integer k. Let α, β be elements of a finite set S. Then an $\alpha\beta$-*ordering* of S is an ordering $\xi_1, \xi_2, \ldots, \xi_n$ of S such that $\xi_1 = \alpha$ and $\xi_n = \beta$. If α, β are vertices of a connected graph G and if $V(G)$ has a k-gradual $\alpha\beta$-ordering, we shall say that G is k-$\alpha\beta$-*orderable*.

Of course, distances between vertices are measured in a particular graph, so that, if α, β are common vertices of two connected graphs G and H under discussion, the distance between α and β in G could be different from the distance between α and β in H: in

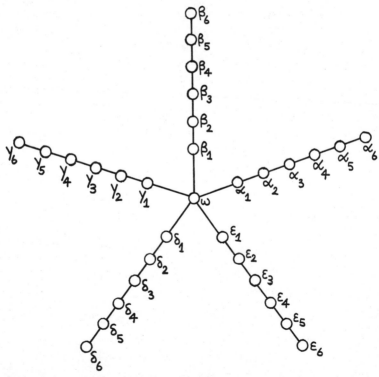

Fig. 18

such a case, we denote these two distances by $d_G(\alpha, \beta)$ and $d_H(\alpha, \beta)$ respectively. For example, if G is the graph of Fig. 10 and H is the subgraph of G depicted in Fig. 9, then $d_G(\theta, \phi) = 1$ but $d_H(\theta, \phi) = 2$. This illustrates the following lemma:

LEMMA 7: *If* α, β *are vertices of a connected subgraph* H *of a connected graph* G, *then* $d_H(\alpha, \beta) \geq d_G(\alpha, \beta)$.

Proof: Since $d_H(\alpha, \beta)$ is the minimum of the lengths of all $\alpha\beta$-paths in H, there must be an $\alpha\beta$-path P in H such that $d_H(\alpha, \beta) = \ell(P)$. Since H is a subgraph of G, it follows that P is an $\alpha\beta$-path in G and so, since $d_G(\alpha, \beta)$ is the minimum of the lengths of all $\alpha\beta$-paths in G, it follows that $d_G(\alpha, \beta) \leq \ell(P) = d_H(\alpha, \beta)$.

Since the definition of a k-gradual sequence of vertices involves distances, which are measured in some graph, it follows that the term "k-gradual" may also have different meanings in different graphs: for instance, the sequence θ, ϕ, γ, β is 1-gradual in the graph of Fig. 10 but not 1-gradual in the subgraph of this graph shown in Fig. 9. For our subsequent discussion, the following corollary of Lemma 7 will be needed:

COROLLARY 7a: *Let* H *be a connected subgraph of a connected graph* G *and* k *be a positive integer. If a sequence on* $V(H)$ *is* k-gradual in H, *then it is* k-gradual in G.

Proof: Let $\xi_1, \xi_2, \ldots, \xi_n$ be a sequence on $V(H)$ which is k-gradual in H. If $n = 1$, the sequence is k-gradual in G since a sequence on $V(G)$ with only one term is automatically considered to be k-gradual in G. Otherwise, for $i = 1, 2, \ldots, n - 1$, we have $d_H(\xi_i, \xi_{i+1}) \leq k$ because our sequence is k-gradual in H and $d_G(\xi_i, \xi_{i+1}) \leq d_H(\xi_i, \xi_{i+1})$ by Lemma 7. Hence $d_G(\xi_i, \xi_{i+1}) \leq k$ for $i = 1, 2, \ldots, n - 1$, i.e., the sequence $\xi_1, \xi_2, \ldots, \xi_n$ is k-gradual in G.

A particularly important and simple kind of graph, which makes its appearance time and again in graph theory, is a *tree*, which may be defined to be a connected graph T such that $T - \lambda$ is disconnected for every edge λ of T. Such a graph has the kind of tree-like structure illustrated by Fig. 19. The following lemma states some elementary properties of trees which will be relevant to the proof of our next main theorem.

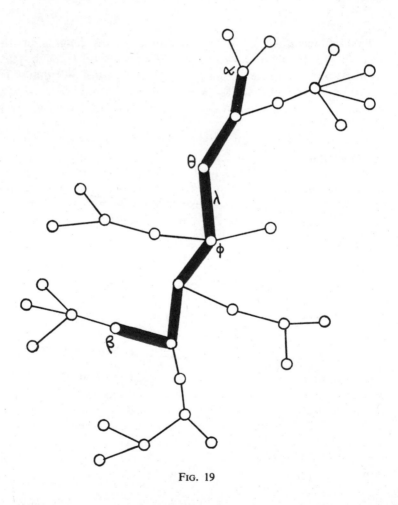

Fig. 19

LEMMA 8: *Let α, β be distinct vertices of a tree T. Then there is one and only one αβ-path in T. If λ is any edge of this αβ-path then*

  (i) *T − λ has exactly two components,*

  (ii) *one of these components includes α and the other includes β,*

  (iii) *one component of T − λ includes one end-vertex of λ and the other component of T − λ includes the other end-vertex of λ.*

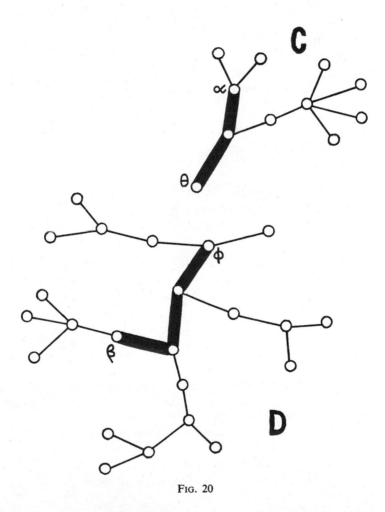

FIG. 20

For instance, Fig. 19 depicts a tree two of whose vertices are labelled α, β, and the unique $\alpha\beta$-path in the tree is the path whose edges are shown as thick edges. One edge in this path is labelled λ; and Fig. 20 depicts the graph $T - \lambda$, which has two components C and D as predicted by Lemma 8(i), and $\alpha \in V(C)$, $\beta \in V(D)$ as predicted by Lemma 8(ii), and C includes the end-vertex θ of λ whilst D includes the end-vertex ϕ of λ, as predicted by Lemma 8(iii). If we replace Fig. 19 by any other diagram depicting a tree T and select any two distinct vertices α, β of this tree, it will again be apparent from the diagram that the assertions of Lemma 8 are valid for this choice of T, α and β; and Lemma 8 seems so intuitively obvious when a few such diagrams have been considered that it seems reasonable to leave its proof as an exercise.

THEOREM 5 (Sekanina [22]): *If G is a connected graph and α, β are distinct vertices of G, then G is 3-$\alpha\beta$-orderable.*

Proof: Our proof will be by induction on the number of edges of G, by showing that Theorem 5 is true for G if it is true for all graphs with fewer edges. More precisely, let us make the following assumptions:

(I) α, β are distinct vertices of a connected graph G.

(II) G' is 3-$\alpha'\beta'$-orderable for every triple G', α', β' such that G' is a connected graph with fewer edges than G and α', β' are distinct vertices of G'.

If, from these assumptions, we can deduce that G is 3-$\alpha\beta$-orderable, then Theorem 5 will be proved by induction.

Suppose, first, that G is not a tree. Since G is connected but not a tree, it follows from the definition of a tree that G has an edge λ_0 such that $G - \lambda_0$ is connected. Since $G - \lambda_0$ is connected and $|E(G - \lambda_0)| < |E(G)|$, it follows from our inductive hypothesis (II) that $G - \lambda_0$ is 3-$\alpha\beta$-orderable. Consequently, there exists an $\alpha\beta$-ordering of $V(G - \lambda_0)$ which is 3-gradual in $G - \lambda_0$. But $V(G - \lambda_0) = V(G)$, and any sequence on this set which is 3-gradual in $G - \lambda_0$ will also by Corollary 7a be 3-gradual in G.

Hence there exists an $\alpha\beta$-ordering of $V(G)$ which is 3-gradual in G, i.e., G is 3-$\alpha\beta$-orderable.

Now suppose that G is a tree. Then, by Lemma 8, there is a unique $\alpha\beta$-path in G and, if we select any edge λ of this path, $G - \lambda$ will have exactly two components, one of which includes α and one end-vertex of λ whilst the other includes β and the other end-vertex of λ. Let C, D be the two components of $G - \lambda$, C being the one which includes α and D being the one which includes β. Let θ be the end-vertex of λ which belongs to $V(C)$ and ϕ be the one which belongs to $V(D)$. We now prove the following two statements:

(i) There is an ordering $\xi_1, \xi_2, \ldots, \xi_r$ of $V(C)$ which is 3-gradual in C and has the properties that $\xi_1 = \alpha$ and ξ_r is either θ or a vertex adjacent to θ in C.

(ii) There is an ordering $\eta_1, \eta_2, \ldots, \eta_s$ of $V(D)$ which is 3-gradual in D and has the properties that $\eta_s = \beta$ and η_1 is either ϕ or a vertex adjacent to ϕ in D.

To prove (i), consider separately the three cases in which (a) $\alpha \neq \theta$, (b) $\alpha = \theta$ and α (*alias* θ) is not the only vertex of C, and (c) $\alpha = \theta$ and α (*alias* θ) is the only vertex of C. Since C is a connected graph with fewer edges than G, our inductive hypothesis (II) tells us in Case (a) that there exists an $\alpha\theta$-ordering of $V(C)$ which is 3-gradual in C, and taking this ordering to be $\xi_1, \ldots, \xi_r$ we have $\xi_1 = \alpha$, $\xi_r = \theta$. In Case (b), since α is not the only vertex of C, the connectedness of C clearly requires that α be incident with at least one edge of C: let σ be a vertex joined to α by an edge of C incident with it. Then by our inductive hypothesis (II) there exists an $\alpha\sigma$-ordering of $V(C)$ which is 3-gradual in C: taking this ordering to be $\xi_1, \ldots, \xi_r$ we have $\xi_1 = \alpha$ and $\xi_r = \sigma$, which is adjacent in C to α and hence (since $\alpha = \theta$) to θ. In Case (c), where α ($= \theta$) is the only vertex of C, take $\xi_1, \ldots, \xi_r$ to be the only possible ordering of $V(C)$, i.e., take $r = 1$ and $\xi_1 = \xi_r = \alpha = \theta$: this ordering of $V(C)$ is trivially 3-gradual in C and has the desired properties $\xi_1 = \alpha$, $\xi_r = \theta$. This completes the proof of (i); and evidently (ii) can be proved in a very similar manner.

The orderings $\xi_1, \ldots, \xi_r$ and $\eta_1, \ldots, \eta_s$ given by (i) and (ii) can be combined to yield a sequence

$$\xi_1, \xi_2, \ldots, \xi_r, \eta_1, \eta_2, \ldots, \eta_s \qquad (16)$$

and we easily see that (16) is an $\alpha\beta$-ordering of $V(G)$ which is 3-gradual in G. In fact, since C, D are the components of $G - \lambda$ and $\xi_1, \ldots, \xi_r$ and $\eta_1, \ldots, \eta_s$ are orderings of $V(C)$, $V(D)$ respectively, it follows that (16) is an ordering of $V(G - \lambda) = V(G)$. Moreover, this ordering of $V(G)$ is an $\alpha\beta$-ordering because $\xi_1 = \alpha$ by (i) and $\eta_s = \beta$ by (ii). To see that (16) is 3-gradual in G, observe, first, that the sequences $\xi_1, \ldots, \xi_r$ and $\eta_1, \ldots, \eta_s$, being 3-gradual in C, D respectively, are by Corollary 7a 3-gradual in G, and, secondly, that $d_G(\xi_r, \eta_1) \leqslant 3$ because ξ_r is equal or adjacent to θ, which is joined by λ to ϕ, which is equal or adjacent to η_1.

Since (16) is a 3-gradual $\alpha\beta$-ordering of $V(G)$, we have now proved that G is 3-$\alpha\beta$-orderable.

THEOREM 6: *Every connected graph is 3-round.*

Proof: A graph with just one vertex is trivially 3-round, and a graph with no vertices (i.e., the empty graph) is 3-round by convention. If G is a connected graph with two or more vertices, then the connectedness of G implies that it must have at least one edge, and if α, β are the vertices joined by some edge of G, then by Theorem 5 there is a 3-gradual $\alpha\beta$-ordering of $V(G)$. Since $d(\beta, \alpha) = 1 < 3$, this 3-gradual ordering is in fact 3-cyclic, and so G is 3-round.

A vertex ξ of a connected graph G is a *cut-vertex* of G if G can be expressed as the union of two subgraphs H and K such that $E(H) \neq \emptyset$, $E(K) \neq \emptyset$ and $V(H) \cap V(K) = \{\xi\}$. For example, if G is the graph of Fig. 21, then ω is a cut-vertex of G because G can be expressed as the union of two subgraphs H, K such that $V(H) = \{\xi_1, \xi_2, \xi_3, \omega\}$, $E(H) = \{\lambda_1, \lambda_2, \lambda_3, \lambda_4, \lambda_5\}$, $V(K) = \{\omega, \xi_4, \xi_5, \xi_6, \xi_7\}$, $E(K) = \{\lambda_6, \lambda_7, \lambda_8, \lambda_9, \lambda_{10}, \lambda_{11}\}$ and these subgraphs satisfy the conditions $E(H) \neq \emptyset, E(K) \neq \emptyset, V(H) \cap V(K) = \{\omega\}$. In the graph of Fig. 9, α is a cut-vertex: there is in

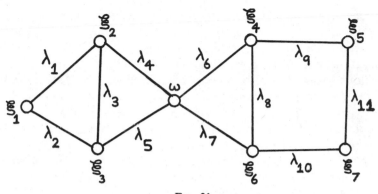

FIG. 21

fact more than one way of expressing this graph as a union of two subgraphs H, K with $E(H) \neq \varnothing$, $E(K) \neq \varnothing$, $V(H) \cap V(K) = \{\alpha\}$, since we could for example choose H, K so that $V(H) = \{\alpha, \theta\}$, $E(H) = \{\lambda\}$, $V(K) = \{\alpha, \beta, \gamma, \phi\}$, $E(K) = \{\mu, \nu, \pi, \rho\}$ or alternatively we could choose them so that $V(H) = \{\alpha, \theta, \phi\}$, $E(H) = \{\lambda, \mu\}$, $V(K) = \{\alpha, \beta, \gamma\}$, $E(K) = \{\nu, \pi, \rho\}$. The central vertex in Fig. 7 is a cut-vertex of the graph, the graph of Fig. 5 has five cut-vertices ξ_2, ξ_3, ξ_4, ξ_5, ξ_6 [for instance, ξ_3 is a cut-vertex because the graph is the union of subgraphs H, K with $V(H) = \{\xi_1, \xi_2, \xi_3\}$, $V(K) = \{\xi_3, \xi_4, \xi_5, \xi_6, \xi_7\}$, $E(H) = \{\lambda_1, \lambda_2\} \neq \varnothing$, $E(K) = \{\lambda_3, \lambda_4, \lambda_5, \lambda_6\} \neq \varnothing$, $V(H) \cap V(K) = \{\xi_3\}$], and in the graph of Fig. 19 every vertex of valency $\geqslant 2$ is a cut-vertex. Thus, intuitively, a cut-vertex of a connected graph is characterised by the property that we can cut the graph into two parts, each containing at least one edge, by cutting through that vertex. A connected graph which has no cut-vertices is said to be *nonseparable*: for example, the graphs in Figs. 1, 2, 3, 4, 8, 10 and 16 are nonseparable.

As we have remarked, the counter-example in Fig. 18 shows that some connected graphs fail to be 2-round, but Sekanina [23] proposed the problem of trying to characterise those which *are* 2-round. Subsequently L. W. Beineke and M. D. Plummer, and independently the present author [15], conjectured that *every non-separable graph is 2-round.* For several years, this very tempting conjecture defeated a considerable number of people who thought about it; but in 1971 H. Fleischner [10, 11] discovered an ingenious proof, too elaborate to be reproduced here. Once this notable theorem had been proved, it was (as several people noticed independently) a not too difficult next step to prove a theorem related to it as Theorem 5 is related to Theorem 6. This theorem asserts that, *if α, β are distinct vertices of a nonseparable graph G, then G is 2-αβ-orderable,* and one method of proof [4] is to deduce it as a corollary to Fleischner's theorem by a trick which consists in constructing, from the given graph G, a new nonseparable graph whose 2-roundness implies the 2-αβ-orderability of G. In view of the important breakthrough achieved by Fleischner, it may be realistic to predict the complete characterisation of 2-round graphs as a possible achievement in graph theory during the next few years.

We have hitherto been viewing 2-roundness, 3-roundness, 4-roundness, etc., as properties of a graph which are weaker than the property of having a Hamiltonian circuit (which, for graphs with at least three vertices, is equivalent to being 1-round). However, there is also another type of relationship between k-roundness and Hamiltonian circuits: a connected graph G with at least three vertices is k-round if and only if a certain related graph G^k has a Hamiltonian circuit, and, more specifically, any ordering of $V(G)$ which is G-k-cyclic (i.e., k-cyclic in G, as opposed to G^k) corresponds in a certain way to a particular Hamiltonian circuit of G^k. The graph G^k is defined by specifying that it has the same vertices as G and that two distinct vertices ξ, η of G are joined by an edge of G^k if and only if $d_G(\xi, \eta) \leqslant k$. For example, if G is the graph of Fig. 17, then G^2 will be as depicted in Fig. 22 and G^3 as in Fig. 23. It is customary to call G^2 the *square* of G, G^3 the *cube* of G and G^k the kth *power* of G: whether or not this nomenclature is fully appropriate might be an arguable question into which we shall not digress. It is easy to see that, if G is a graph with at least three vertices, any G-k-cyclic ordering of $V(G)$ gives rise to a Hamiltonian circuit of G^k and any Hamiltonian circuit of G^k gives rise to a G-k-cyclic ordering of $V(G)$: for example, if G is the graph of

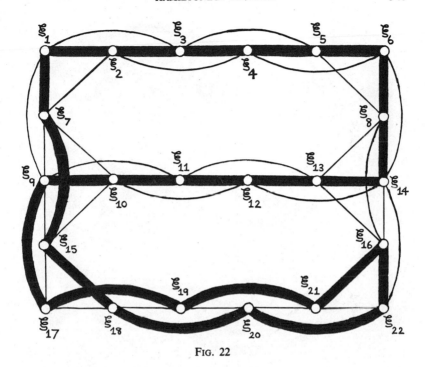

Fig. 22

Fig. 17, then the G-2-cyclic ordering (15) of $V(G)$ gives rise to the Hamiltonian circuit of G^2 whose edges are the thick edges in Fig. 22 and conversely, given this Hamiltonian circuit of G^2, we could use it to construct the G-2-cyclic ordering (15) of $V(G)$ by listing the vertices in the order of their occurrence as we go round the circuit starting at ξ_1. (Of course, if, for example, we started at ξ_5 and went round the circuit in the opposite direction, this would yield a different G-2-cyclic ordering

$$\xi_5, \xi_4, \xi_3, \xi_2, \xi_1, \xi_7, \xi_{15}, \xi_{18}, \xi_{20}, \xi_{22}, \xi_{16}, \xi_{21},$$
$$\xi_{19}, \xi_{17}, \xi_9, \xi_{10}, \xi_{11}, \xi_{12}, \xi_{13}, \xi_{14}, \xi_8, \xi_6$$

of $V(G)$; so there are strictly speaking *several* k-cyclic orderings of the set of vertices of a graph associated with any Hamiltonian circuit of its kth power, but this in no way conflicts with what we

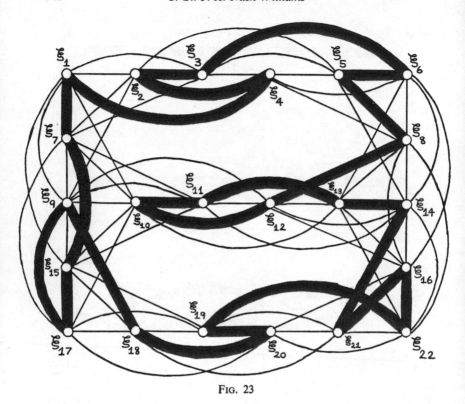

FIG. 23

have been saying.) Fig. 23 exhibits by means of thick edges the Hamiltonian circuit of G^3 associated with the G-3-cyclic ordering

$$\xi_1, \ \xi_4, \ \xi_2, \ \xi_3, \ \xi_6, \ \xi_5, \ \xi_8, \ \xi_{12}, \ \xi_{10}, \ \xi_{11}, \ \xi_{13},$$
$$\xi_{14}, \ \xi_{21}, \ \xi_{16}, \ \xi_{22}, \ \xi_{19}, \ \xi_{20}, \ \xi_{18}, \ \xi_9, \ \xi_{17}, \ \xi_{15}, \ \xi_7$$

of $V(G)$, where G is the graph of Fig. 17. These examples should make it clear that k-roundness of a connected graph G with at least 3 vertices is essentially the same thing as roundness of G^k, and so Theorem 6 and Fleischner's theorem can be expressed respectively in the forms *the cube of every connected graph with at least three vertices is round* and *the square of every nonseparable graph with at least three vertices is round.*

5. TOUGHNESS AND HAMILTONIAN CIRCUITS

If X is a subset of $V(G)$, then $G - X$ will denote the graph obtained from G by removing the vertices in X and all edges incident with them: for example, if G is the graph of Fig. 24 and X is the set of "square" vertices, then $G - X$ is as shown in Fig. 25.

The number of connected components of G will be denoted by $c(G)$.

In looking for possible sufficient conditions for graphs to be round, one may tend to notice that some of the most easily found examples of tortuous graphs are, in a certain sense, "structurally weak," or "badly connected" graphs. For example, disconnected graphs are obviously tortuous. Again, connected graphs with at least one cut-vertex are obviously tortuous, and these graphs might be considered as being "structurally weak" or "badly connected" in the sense that "one has only to cut through one vertex in order to disconnect them." Again, a graph like that of Fig. 27 is easily seen to be tortuous, and this graph is "structurally weak" in the sense that three parts of it are held together only at the two vertices α, β: more mathematically speaking, $G - \{\alpha, \beta\}$ is a disconnected graph with three connected components, where G denotes the graph in Fig. 27. We shall now prove a lemma and corollary based on the idea that "structurally weak graphs tend to be tortuous."

LEMMA 9: *If G is round, then $c(G - X) \leqslant |X|$ for every non-empty subset X of $V(G)$.*

Proof: Since G is round, we can select a Hamiltonian circuit C of G. Let X be a non-empty subset of $V(G)$. Clearly the connected components of $C - X$ are disjoint paths $A_1, A_2, \ldots, A_r$, say, and, if we start at some vertex (α, say) in X and travel round C in a selected direction until we arrive back at α, we shall pass along each A_i once. Let ξ_i be the next vertex of C encountered after we have passed through all the vertices of A_i in this journey round C. Then clearly $\xi_1, \ldots, \xi_r$ are distinct elements of X, and hence

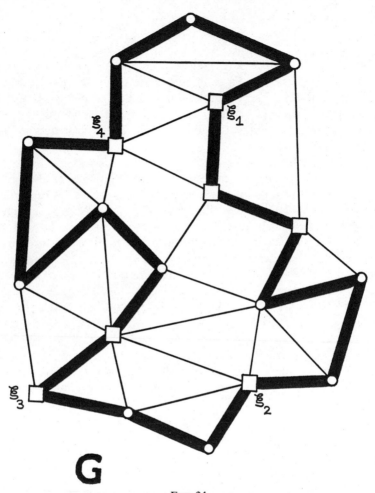

G

Fig. 24

$|X| \geqslant r$. Moreover, $A_1, \ldots, A_r$ include between them all the vertices of $G - X$ and, since each A_i is a connected subgraph of $G - X$, its vertices must all belong to the same connected component of $G - X$. Hence $G - X$ can have at most r connected components, i.e., $c(G - X) \leqslant r$. We have thus proved that

$$c(G - X) \leqslant r \leqslant |X|.$$

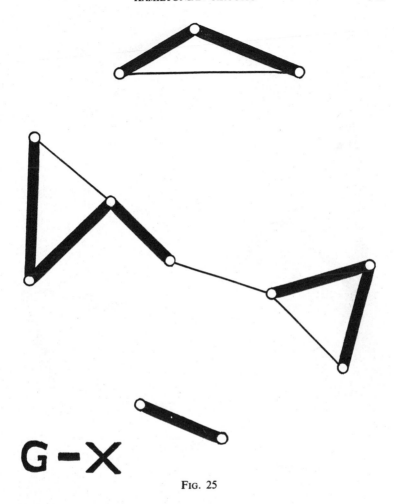

FIG. 25

This proof is illustrated by Figs. 24, 25 and 26, which represent G, $G - X$ and $C - X$ respectively, the elements of X being the "square" vertices in Fig. 24. Edges of C are shown as thick edges in these diagrams.

Another way of writing Lemma 9 is:

COROLLARY 9a: *If $c(G - X) > |X|$ for some non-empty subset X of $V(G)$, then G is tortuous.*

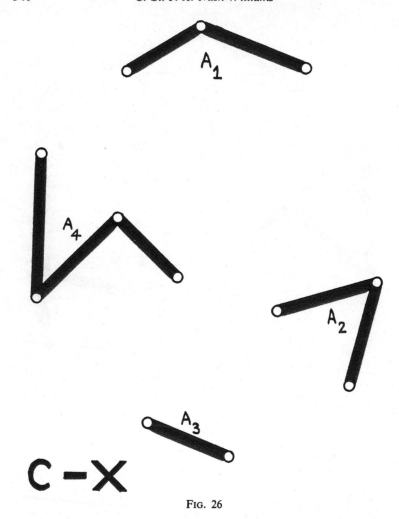

FIG. 26

For example, G is tortuous if (as in Fig. 21) $G - \{\omega\}$ has at least two connected components for some vertex ω of G or if (as in Fig. 27) $G - \{\alpha, \beta\}$ has at least three connected components for some pair of vertices α, β of G. In general, graphs which become split up into a relatively large number of connected components by the removal of a relatively small number of vertices (and the edges incident with these vertices) may be thought of as being

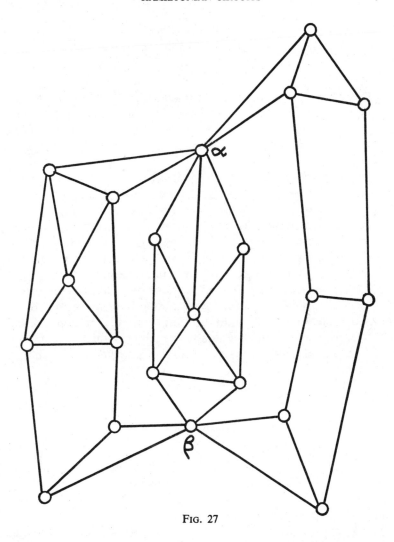

FIG. 27

"structurally weak" and Corollary 9a says, roughly speaking, that graphs with a certain degree of "structural weakness" are necessarily tortuous. One might ask whether, on the other hand, graphs with a certain degree of "structural strength" or "toughness" are necessarily round, and in this direction Chvátal [6] has proposed the following conjecture.

CONJECTURE 3: *If $|V(G)| \geqslant 3$ and $c(G - X) < \frac{2}{3}|X|$ for every subset X of $V(G)$ such that $G - X$ is disconnected, then G is round.*

[The words "such that $G - X$ is disconnected" could not reasonably be left out in Conjecture 3: if we required $c(G - X)$ to be less than $\frac{2}{3}|X|$ for *every* subset X of $V(G)$ then we would be requiring it to be less than $\frac{2}{3}$ when $|X| = 1$ and to be less than 0 when $|X| = 0$.]

As Chvátal pointed out in [6], it is a tolerably easy exercise to prove that, if G is any nonseparable graph, then $c(G^2 - X) \leqslant \frac{1}{2}|X|$ for every subset X of $V(G^2)$ [$= V(G)$] such that $G^2 - X$ is disconnected. Hence the square of any nonseparable graph with at least three vertices satisfies the hypotheses of Conjecture 3 and so, if we could prove Conjecture 3, then this result would contain the statement that the square of every nonseparable graph with at least three vertices is round; and we have seen that this statement is essentially a reformulation of Fleischner's theorem that every nonseparable graph is 2-round. So Conjecture 3, if true, must be a sufficiently deep result to contain Fleischner's theorem, and indeed it would evidently go considerably beyond Fleischner's theorem.

6. HAMILTONIAN CIRCUITS IN PLANAR GRAPHS

Informally speaking, a *planar* graph is a graph drawn in the plane so that no two edges intersect and no edge passes *through* any vertex (although of course the points at each *end* of an edge must be vertices). For example, if Figures 1, 2, 4, 5, 9, 10, 17, 18, 19, 20, 21, 24, 25, 26, 27, 28, 29 and 30 are interpreted as representing graphs drawn in the plane, then all of these graphs are planar; but Figs. 3, 7, 8, 16, 22 and 23 represent graphs drawn in the plane with some pairs of intersecting edges, and such graphs are not planar. These examples should adequately indicate what we mean by a "planar" graph, but for those desiring a more precise and more technical definition we can define a *planar* graph* to be a graph G such that

(i) the vertices of G are points of R^2, where R^2 denotes the Euclidean plane,

(ii) the edges of G are disjoint subsets of $R^2 \backslash V(G)$,

(iii) for each edge λ of G, there exists a continuous one-to-one mapping f_λ of the closed interval $[0, 1]$ into R^2 such that λ is the image under f_λ of the open interval $(0, 1)$ and $f_\lambda(0), f_\lambda(1)$ are the vertices joined by λ in G.

Any planar graph G can be re-drawn, if we so wish, on the surface of a sphere. All we have to do is cut out from the plane a circular disc which contains all the vertices and edges of G and, assuming it to be made of sufficiently elastic material, paste it onto the surface of the sphere, and we will then have G drawn on the surface of the sphere.

A graph G is sometimes said to be *r-connected* (where r is a positive integer) if $G - X$ is connected for every subset X of $V(G)$ such that $|X| < r$. Of course, this includes saying that $G - X$ is connected when $X = \emptyset$, i.e., that G itself is connected. Thus an r-connected graph is a connected graph which cannot be rendered disconnected by removing *less* than r vertices. (When we speak of "removing" vertices from a graph, it is always understood that the edges incident with those vertices are removed as well: see the definition of $G - X$ in Section 5.)

An elementary property of r-connected graphs is stated in the following lemma:

LEMMA 10: *If an r-connected graph has at least $r + 1$ vertices, then all of its vertices have valency $\geqslant r$.*

Proof: We will show that, if a graph G has at least $r + 1$ vertices

---

*It might, in terms of customary usage, be more accurate to call the type of graph thus defined a *plane graph* and to call any graph which is isomorphic to a plane graph *planar*. However, we shall in this section adhere to the definition given above, since we shall have no particular reason to concern ourselves with graphs isomorphic to plane graphs other than plane graphs themselves.

and some vertex α of G has valency less than r, then G is not r-connected. In fact, if in these circumstances N denotes the set of vertices adjacent to α in G, then $|N| = v(\alpha) < r$ and so G will certainly fail to be r-connected if $G - N$ is disconnected. But $G - N$ is disconnected because (i) α is a vertex of $G - N$ which is not adjacent in $G - N$ to any other vertex of $G - N$ and (ii) $|V(G - N)| > 1$ since $|V(G - N)| = |V(G)| - |N|$, $|V(G)| \geqslant r + 1, |N| < r$.

Tutte [24] has proved the following theorem:

THEOREM 7: *Every 4-connected planar graph with at least three vertices has a Hamiltonian circuit.*

The proof, which may be found in [18] or [24], involves a complicated and ingenious argument by induction on the number of edges of the graph. Theorem 7 generalized a previous result of Whitney [25], which was in fact the special case of Theorem 7 in which the theorem is restricted to 4-connected planar graphs all of whose faces have valency 3, the terms "face" and "valency of a face" being defined later in this section.

In trying to learn more about which graphs have Hamiltonian circuits, it may be useful to have, not only positive results like Theorems 1, 2 and 7 which say that certain graphs have Hamiltonian circuits, but also some examples of graphs which are known *not* to possess them other than rather obvious examples such as those provided by Corollary 9a. An ingenious device which yields a few such examples (including that of Fig. 4) has recently been found in the study of planar graphs, and we conclude with a brief description of it.

Observe, first, that a planar graph divides the rest of the plane (i.e., that part of the plane not occupied by vertices and edges of the graph) into a number of regions which we shall call *faces* of the graph: for instance, the graph of Fig. 28 divides the rest of the plane into 22 regions $I_1, I_2, \ldots, I_9, J_1, J_2, \ldots, J_{13}$, and these 22 regions are the faces of the graph. Clearly one of the faces of a planar graph will be an unbounded region of the plane, extending to infinity in all directions, and the remaining faces will be

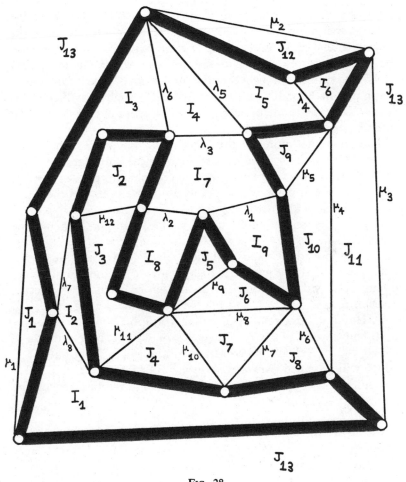

FIG. 28

bounded regions: in Fig. 28 the unbounded face is J_{13} and the other faces are bounded. Thus, informally speaking, the unbounded face is the region "outside the graph." If we take a planar graph and re-draw it (as described above) on the surface of a sphere, then the graph thus re-drawn will divide the rest of the surface of the sphere into regions which might be called "faces" of the re-drawn graph, and these faces will closely correspond to the faces of the graph as originally drawn in the plane, but with the difference that all faces on the sphere are bounded regions, including the one corresponding to the unbounded face in the plane. Thus the distinction between the unbounded face of a planar graph and its bounded faces is, in a sense, a slightly artificial one, since it disappears when the graph is re-drawn on a sphere and all faces become bounded.

The above definition of the faces of a planar graph has been given informally: in more precise topological language, the faces of a planar graph G are the connected components of the subspace of the Euclidean plane whose points are those points of the plane which belong neither to $V(G)$ nor to any edge of G.

If C is a Hamiltonian circuit of a planar graph G, then some faces of G lie inside C and will be called C-internal faces, whilst others lie outside C and will be called C-external faces: for instance, if G is the graph in Fig. 28 and C is the Hamiltonian circuit indicated by the thick edges in Fig. 28, then the C-internal faces of G are $I_1, \ldots, I_9$ and the C-external faces are $J_1, \ldots, J_{13}$.

A Hamiltonian circuit C of a planar graph G divides the edges of G into *three* categories, viz., edges of C, edges lying inside C which will be called C-internal edges, and edges lying outside C which will be called C-external edges: for example, if G and C are as depicted in Fig. 28, then the edges of C are the thick edges, the C-internal edges of G are $\lambda_1, \ldots, \lambda_8$ and the C-external edges of G are $\mu_1, \ldots, \mu_{12}$.

In the example of Fig. 28, we notice that the number of C-internal edges is one less than the number of C-internal faces and the number of C-external edges is one less than the number of C-external faces. This phenomenon is in fact not peculiar to this particular example, but is a consequence of the following general lemma:

LEMMA 11. *If C is a Hamiltonian circuit of a planar graph G and if G has r C-internal faces and s C-external faces, then G has $r - 1$ C-internal edges and $s - 1$ C-external edges.*

To indicate why this is true, we begin by noticing that, if C is a Hamiltonian circuit of a planar graph G, then the C-internal faces form a somewhat tree-like structure, as may be seen in Fig. 28, and the C-external faces form another somewhat tree-like structure, also discernible in Fig. 28. In fact, this statement can be made precise by saying how we can actually associate a tree with the C-internal faces and another with the C-external faces. Our next lemma does this for the C-internal faces.

LEMMA 12: *Let C be a Hamiltonian circuit of a planar graph G. Let the C-internal faces of G be $I_1, \ldots, I_r$ and the C-internal edges of G be $\lambda_1, \ldots, \lambda_y$ and let $T_i(G, C)$ be a graph with r vertices $I_1', \ldots, I_r'$ and y edges $\lambda_1', \ldots, \lambda_y'$ such that, for $k = 1, \ldots, y$, the vertices joined by λ_k' in $T_i(G, C)$ are those corresponding to the C-internal faces on opposite sides of λ_k. (By the vertex of $T_i(G, C)$ "corresponding" to a C-internal face I_α, we mean the vertex I_α'.) Then $T_i(G, C)$ is a tree.*

For instance, if G and C are as in Fig. 28, then G has C-internal faces $I_1, \ldots, I_9$ and C-internal edges $\lambda_1, \ldots, \lambda_8$ and so $T_i(G, C)$ must have nine vertices $I_1', \ldots, I_9'$ and eight edges $\lambda_1', \ldots, \lambda_8'$: since the C-internal faces on opposite sides of λ_1 are I_7 and I_9, λ_1' must join the corresponding vertices I_7' and I_9' in $T_i(G, C)$ and since the C-internal faces on opposite sides of λ_2 are I_7 and I_8, λ_2' must join the corresponding vertices I_7' and I_8', and so forth. Thus $T_i(G, C)$ will be as in Fig. 29.

We shall not actually write out a proof of Lemma 12: to gain insight as to why this lemma is true, the reader is recommended to draw a number of diagrams like Fig. 28 which depict a planar graph G and a Hamiltonian circuit C of G, and then construct the graph $T_i(G, C)$ associated with each diagram. It will in all cases turn out to be a tree, and, if he needs further convincing, the reader is recommended to spend a little time deliberately trying to draw a pair G, C for which $T_i(G, C)$ is *not* a tree.

Similarly, by studying illustrative diagrams, one can easily convince oneself of the truth of the following lemma:

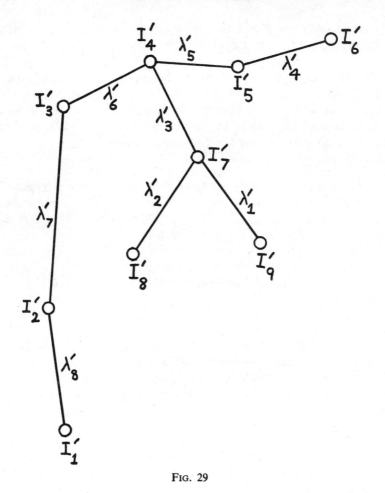

FIG. 29

LEMMA 13: *Let C be a Hamiltonian circuit of a planar graph G.
Let the C-external faces of G be $J_1, \ldots, J_s$ and the C-external
edges of G be $\mu_1, \ldots, \mu_z$ and let $T_e(G, C)$ be a graph with s vertices
$J'_1, \ldots, J'_s$ and z edges $\mu'_1, \ldots, \mu'_z$ such that, for $l = 1, \ldots, z$, the
vertices joined by μ'_l in $T_e(G, C)$ are those corresponding to the
C-external faces on opposite sides of μ_l. (By the vertex of $T_e(G, C)$*

"corresponding" to a *C-external face* J_β, we mean the vertex J'_β.) Then $T_e(G, C)$ is a tree.

If G and C are as in Fig. 28, then $T_e(G, C)$ will be as depicted in Fig. 30.

It is not particularly surprising that Lemma 13, which describes what happens outside C, bears such a close resemblance to Lemma 12, which describes what happens inside C. The distinction between "inside" and "outside" would disappear if we re-

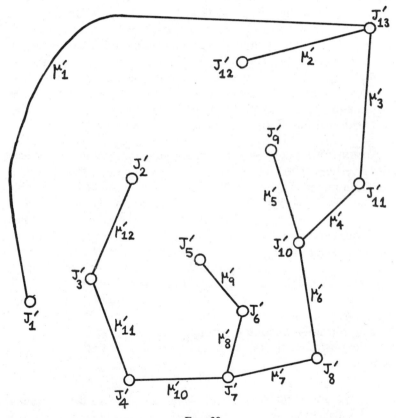

FIG. 30

draw G on the surface of a sphere, when C would become a circuit on the sphere's surface and, although this circuit would divide the sphere's surface into two regions, we could not now identify one of them as being "outside" the circuit and the other as being "inside."

We are now in a position to prove Lemma 11. Let C be a Hamiltonian circuit of a planar graph G. Then the definition of $T_i(G, C)$ implies that the number of C-internal faces of G is equal to the number of vertices of $T_i(G, C)$ and the number of C-internal edges of G is equal to the number of edges of $T_i(G, C)$. But $T_i(G, C)$ is by Lemma 12 a tree, and it is a well-known elementary fact that $|E(T)| = |V(T)| - 1$ *for every non-empty tree* T: the reader can readily convince himself of this fact by examining examples of trees, and a proof by induction is fairly easy to give. Hence $|E(T_i(G, C))| = |V(T_i(G, C))| - 1$, and so the number of C-internal edges of G is one less than the number of C-internal faces. In a similar way, it follows from Lemma 13 that the number of C-external edges of G is one less than the number of C-external faces; and Lemma 11 is proved.

DEFINITION: Let G be a nonseparable planar graph with at least three vertices. Then we define the *valency* $v(F)$ of a face F of G to be the number of edges of G in the boundary of F. For example, in the graph of Fig. 28, $v(J_7) = 3$ since the boundary of J_7 contains three edges μ_7, μ_8, μ_{10} of the graph, and $v(J_{10}) = 4$ since the boundary of J_{10} contains four edges, viz., μ_4, μ_5, μ_6 and one thick edge, and $v(J_{13}) = 5$ since the boundary of J_{13} contains five edges, viz., μ_1, μ_2, μ_3 and two thick edges. These five edges are in the boundary which bounds J_{13} on the *inside*; and in general the valency of the unbounded face of a nonseparable planar graph with three or more vertices will be the number of edges in the circuit forming the outer perimeter of the graph, because this circuit constitutes the only boundary of the unbounded face of the graph.

LEMMA 14: *Let C be a Hamiltonian circuit of a planar graph G. Let $I_1, \ldots, I_r$ be the C-internal faces of G and $J_1, \ldots, J_s$ be the*

C-external faces of G. Then

$$(v(I_1) - 2) + (v(I_2) - 2) + \cdots + (v(I_r) - 2)$$

$$= (v(J_1) - 2) + (v(J_2) - 2) + \cdots + (v(J_s) - 2). \quad (17)$$

Proof: The sum $v(I_1) + v(I_2) + \cdots + v(I_r)$ is obtained by considering the C-internal faces one by one and, for each such face, counting the edges in its boundary. Each C-internal edge gets counted twice in this process because it is in the boundary of two C-internal faces, and each edge of C gets counted once because it is in the boundary of one C-internal face. Hence $v(I_1) + \cdots + v(I_r) = 2y + |E(C)|$, where y is the number of C-internal edges. But $y = r - 1$ by Lemma 11, and hence $v(I_1) + \cdots + v(I_r) = 2(r - 1) + |E(C)|$, i.e., $v(I_1) + \cdots + v(I_r) - 2r = |E(C)| - 2$, i.e.,

$$(v(I_1) - 2) + (v(I_2) - 2) + \cdots + (v(I_r) - 2)$$

$$= |E(C)| - 2. \quad (18)$$

In a similar way, the number of C-external edges of G is $s - 1$ by Lemma 11, and each C-external edge is in the boundary of two C-external faces whilst each edge of C is in the boundary of one such face, so that $v(J_1) + \cdots + v(J_s)$ counts each of the $s - 1$ C-external edges twice and each edge of C once. Hence $v(J_1) + \cdots + v(J_s) = 2(s - 1) + |E(C)|$, i.e.,

$$(v(J_1) - 2) + (v(J_2) - 2) + \cdots + (v(J_s) - 2)$$

$$= |E(C)| - 2. \quad (19)$$

Now (17) follows from (18) and (19).

We have now almost completed the proof of the following elegant theorem of Grinberg [26, 21]:

THEOREM 8: *Suppose that a nonseparable planar graph G has one and only one face whose valency is NOT one of the numbers 5, 8, 11, 14, 17, 20, 23, 26, ... Then G has no Hamiltonian circuit.*

Proof: Suppose that G has a Hamiltonian circuit C. Then, if $I_1, \ldots, I_r$ are the C-internal faces of G and $J_1, \ldots, J_s$ are its C-external faces, the valencies of these faces must by Lemma 14 satisfy (17). It follows from (17) that, if all but one of the numbers $v(I_1) - 2, v(I_2) - 2, \ldots, v(I_r) - 2, v(J_1) - 2, v(J_2) - 2, \ldots, v(J_s) - 2$ are divisible by 3, then the remaining one must also be divisible by 3. In other words, if the valencies of all but one of the faces of G belong to the set $\{5, 8, 11, 14, 17, \ldots\}$, then the valency of the remaining face must also belong to this set, i.e., G cannot have just one face whose valency is not one of $5, 8, 11, 14, 17, \ldots$. But, by hypothesis, G *has* just one such face: so assuming G to have a Hamiltonian circuit leads to a contradiction and Theorem 8 is proved.

The nonseparable planar graph of Fig. 4 has one and only one face whose valency is not one of $5, 8, 11, 14, 17, \ldots$: in fact, the unbounded face of this graph has valency 9 since its boundary contains the nine edges in the circuit running round the outside of the graph, but all of its bounded faces have valencies of 5 or 8. Hence, by Theorem 8, the graph of Fig. 4 has no Hamiltonian circuit. This example is due to Grinberg and appeared in [26] and [21].

REFERENCES

Considerations of space and time have both precluded any attempt to provide an exhaustive bibliography of the subject. The following references are confined to publications referred to above and two survey articles [16, 17]:

1. Bondy, J. A., "Properties of graphs with constraints on degrees", *Studia Sci. Math. Hungar.*, **4** (1969), 473–475.

2. Capobianco, M., J. B. Frechen and M. Krolik (editors), "Recent trends in graph theory", *Proc. First New York City Graph Theory Conference* in *Lecture Notes in Mathematics*, **186**, Berlin, Heidelberg and New York: Springer-Verlag, 1971.

3. Chartrand, G., and S. F. Kapoor (editors), "The many facets of graph theory", *Proc. Conference at Western Michigan University, November*

1968, in *Lecture Notes in Mathematics*, **110**, Berlin, Heidelberg and New York: Springer-Verlag, 1969.

4. Chartrand, G., A. M. Hobbs, H. A. Jung, S. F. Kapoor, and C. St. J. A. Nash-Williams, "The square of a block is Hamiltonian connected", *J. Combinatorial Theory, Ser. B.*, **16** (1974), 290–292.

5. Chvátal, V., "On Hamilton's ideals", *J. Combinatorial Theory, Ser. B*, **12** (1972), 163–168.

6. ——, "Tough graphs and Hamiltonian circuits", *Discrete Math.*, **5** (1973), 215–228.

7. Dirac, G. A., "Some theorems on abstract graphs", *Proc. London Math. Soc.*, **2** (1952), 69–81.

8. Erdös, P., and G. Katona (editors), "Theory of graphs", *Proc. symposium at Tihany, Hungary*, Budapest: Publishing House of the Hungarian Academy of Sciences; New York: Academic Press, 1968.

9. Fiedler, M. (editor), "Theory of graphs and its applications", *Proc. Symposium held in Smolenice in June 1963*; Prague: Czechoslovak Academy of Sciences, 1964.

10. Fleischner, H., "On spanning subgraphs of a connected bridgeless graph and their application to DT-graphs", *J. Combinatorial Theory, Ser. B.*, **16** (1974), 17–28.

11. ——, "The square of every two-connected graph is Hamiltonian", *J. Combinatorial Theory, Ser. B.*, **16** (1974), 29–34.

12. Ghouila-Houri, A., "Une condition suffisante d'existence d'un circuit Hamiltonien", *C. R. Acad. Sci. Paris*, **251** (1960), 495–497.

13. Harary, F., *Graph Theory*, Reading, Mass.: Addison-Wesley, 1969.

14. Nash-Williams, C. St. J. A., *Unsolved Problem*, No. 47, on page 366 of reference 8 above.

15. ——, *Unsolved Problem*, No. 48, on page 367 of reference 8 above.

16. ——, *Hamiltonian circuits in graphs and digraphs*, on pages 237–243 of reference 3 above.

17. ——, *Hamiltonian arcs and circuits*, on pages 197–210 of reference 2 above.

18. Ore, O., "The four-color problem", in *Pure and Applied Mathematics*, **27**, New York and London: Academic Press, 1967.

19. Pósa, L., "A theorem concerning Hamiltonian lines", *Magyar Tud. Akad. Mat. Fiz. Oszt. Kozl.*, **7** (1962), 225–226.

20. Sachs, H., H.-J. Vosz, and H. Walther (editors), "Beiträge zur Graphentheorie, vorgetragen auf dem internationalen Kolloquium in Manebach (DDR), 9.-12.Mai 1967", Leipzig: B. G. Teubner, 1968.

21. Sachs, H., Ein von Kozyrev und Grinberg angegebener nichthamiltonischer kubischer planarer Graph, on pages 127–130 of reference 20 above.

22. Sekanina, M., "On an ordering of the set of vertices of a connected graph", Spisy Prirod. Fak. Univ. Brno, 412 (1960), 137–141.

23. ——, Unsolved Problem, No. 28, on page 164 of reference 9 above.

24. Tutte, W. T., "A theorem on planar graphs", Trans. Amer. Math. Soc., 82 (1956), 99–116.

25. Whitney, H., "A theorem on graphs", Ann. of Math., 32 (1931), 378–390.

26. Grinberg, È. Ja., "Plane homogeneous graphs of degree three without Hamiltonian circuits" (Russian, Latvian and English summaries) Latvian Math. Yearbook, 4 (1968), 51–58 (Izdat. "Zinatne", Riga, 1968).

CHROMIALS

W. T. Tutte

In the beginning was the Four Colour Problem. It was the problem of proving that for every possible planar map, the regions, faces or countries can be coloured in not more than four distinct colours so that no two of the same colour have a common frontier line, or of finding a counter-example.

Many there were who sought to solve the problem, and all their methods were but one method. It is a method that is practiced even unto this day. One assumes a map M such that all maps with fewer faces can be four-coloured, and one tries to deduce from this information that M is itself four-colourable. It is called the "qualitative method" by Birkhoff and Lewis in their great work *Chromatic Polynomials* [4]. But the workings of this method, and the fruits thereof, are they not written in the book of Ore? [10].

In a paper of 1912, G. D. Birkhoff called for a quantitative method. "Let us not," said he in effect, "be content with the distinction between four-colourable and not four-colourable. For each map M there is an integer $P(M, 4)$ which is the number of ways of four-colouring it. Let us study the properties of this function of a general map. And while we are about it, let us generalize the function to other numbers of colours than four. Let us study the function $P(M, \lambda)$, which is the number of ways of colouring the map M in λ colours" [3]. At this stage, we should try to be precise in our definitions. We suppose given a set of λ "colours." The integers from 1 to λ will do very nicely. A λ-colouring is an assignment of exactly one of these colours to each face so that no two faces with a common boundary have the same colour. Not all the λ colours need actually be used, and a (non-identical) permutation of the colours actually used is considered to give a new λ-colouring.

Birkhoff found that for each map M the function $P(M, \lambda)$ was a polynomial in λ whose degree was the number of faces of M. Wherefore this function was called the chromatic polynomial of M, or by a recent abbreviation the chromial of M.

Clearly, Birkhoff hoped to gain deeper insight into colouring problems by switching to the quantitative method and a general λ. He was encouraged by finding some simple general formulae connecting the chromials of three or more closely related maps. These made fairly easy the computation of the chromial of a reasonably simple map, but they made no detectable breach in the defences of the Four Colour Problem.

In 1932, Hassler Whitney showed that the notion of a chromatic polynomial was applicable also in the case of vertex-colourings of graphs. In this case, one has a graph instead of a map, and the graph need not be planar. Again one has a set of λ colours, and a vertex-colouring is an assignment of exactly one colour to each vertex, subject to the restriction that the two ends of an edge must receive two distinct colours. The number of λ-colourings of a graph G is denoted by $P(G, \lambda)$. This function turns out to be a polynomial in λ, and if G is loopless, its degree is the number of vertices. If G has a loop then no λ-colouring is possible and $P(G, \lambda)$ is identically zero. In either case $P(G, \lambda)$ is called the chromatic polynomial or chromial of G [13].

Whitney's paper is fundamental in the theory of the chromials of general graphs. In it he even contemplates a further generalization involving two colour-numbers instead of one, —to what is now often called a dichromatic polynomial. But this generalization goes beyond the scope of the present article.

The chromials of Whitney include those of Birkhoff. A map M can be replaced by its dual graph G. This has one vertex for each face of M, and two vertices are joined by an edge if and only if the corresponding faces of M have a common boundary line. Then evidently $P(M, \lambda) = P(G, \lambda)$. The graph G can be drawn in the plane without crossings by the device of taking each vertex to be a point inside the corresponding face, and making appropriate joins across face-boundaries. Usually the exterior of a planar map is counted as one of the faces, and this too has its corresponding vertex of G.

Because of the correspondence between a map M and its dual graph G it is possible, and it is now becoming customary, to state the Four Colour Conjecture in the following form: for any loopless planar graph G, $P(G, 4) > 0$. It is to be noted that a map M, as a map is usually defined, does not give rise to a dual graph with a loop; no frontier line separates a face of M from itself.

A loopless connected graph G drawn in the plane defines a map N whose frontier lines are the edges of G. It may happen that each face of N (including the outer one) is triangular, that is, bounded by a simple closed curve made up of exactly three edges of G. In this case, we call N a triangulation of the plane. If N is not a triangulation, it can be made one by adding new edges to G. From this observation it is deduced that if the Four Colour Theorem is true for the graphs defining triangulations, then it is true in general. Accordingly, it is the graphs of this kind that receive most attention in the literature. In what follows we do not distinguish between a planar triangulation and its defining graph. When we speak of a colouring of such a figure, we shall always mean a vertex-colouring. In the dual theory of face-colourings, the triangulations are transformed into the trivalent maps, in which each vertex lies on the boundary of exactly three faces.

In 1946, G. D. Birkhoff and D. C. Lewis published a long paper called *Chromatic Polynomials* [4], which ever since has been the chief source of information and inspiration for workers in this field. Its theory is that of trivalent maps and face-colourings, but in describing its results we shall translate them into the terminology of triangulations and vertex-colourings. Much of the paper is concerned with the problem of calculating the chromial of a given triangulation. Their attack on this problem is based on three very simple recursion formulae. We state these below, the first two in generalized form for the sake of later analogies.

I. Consider a triangulation T with a separating digon. Such a triangulation is shown in Figure 1, the digon having edges A and B. If we delete from T the edges and vertices outside the digon, and then delete one edge of the digon, we obtain a simpler triangulation T_1. Similarly, if we delete the edges and vertices inside the digon, and then delete one edge of the digon, we obtain a triangulation T_2. It is easily seen that

$$P(T, \lambda) = \frac{P(T_1, \lambda)P(T_2, \lambda)}{\lambda(\lambda - 1)} \qquad (\lambda > 1). \qquad (1)$$

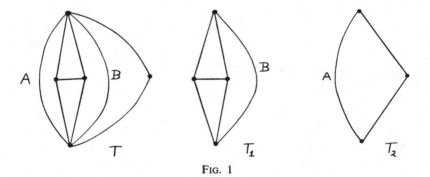

FIG. 1

II. Consider a triangulation T with a separating triangle (abc in Figure 2). If we delete from T the edges and vertices inside the triangle, we obtain a simpler triangulation T_1. Similarly, if we delete the edges and vertices outside the triangle we obtain a triangulation T_2. It is clear that

$$P(T, \lambda) = \frac{P(T_1, \lambda)P(T_2, \lambda)}{\lambda(\lambda - 1)(\lambda - 2)} \qquad (\lambda > 2). \qquad (2)$$

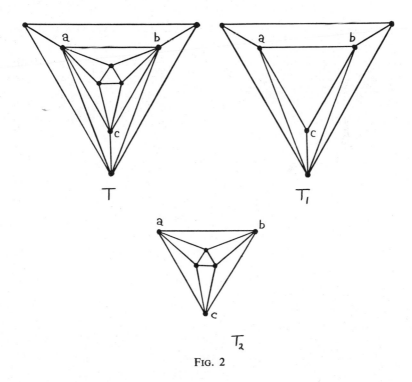

FIG. 2

III. Consider a triangulation T in which two triangles with a common edge $A = ac$ form the inside of a quadrilateral $abcd$, (Figure 3). From T we can form a triangulation $\theta_A(T)$ by "twisting" A, that is, replacing A by the other diagonal of the quadrilateral. If T is without separating digons, we can form a triangulation $\phi_A(T)$ by contracting A to a single vertex and correspondingly contracting each of the two triangles to a single segment. The analogous operation on $\theta_A(T)$ yields a triangulation $\psi_A(T)$. It is found that

$$P(T, \lambda) + P(\phi_A(T), \lambda) = P(\theta_A(T), \lambda) + P(\psi_A(T), \lambda). \quad (3)$$

This identity is most easily proved by considering the map N, with one quadrangular face, obtained from T by erasing the edge A.

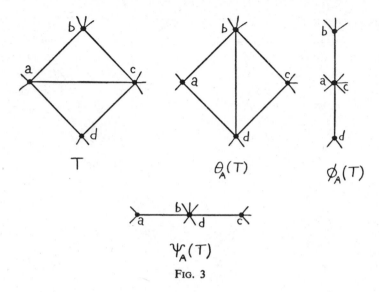

$$\underset{A}{\theta}(T)$$

$$\underset{A}{\phi}(T)$$

$$\underset{A}{\psi}(T)$$

FIG. 3

$P(T, \lambda)$ can be interpreted as the number of λ-colourings of N in which a and c have different colours, and $P(\phi_A(T), \lambda)$ as the number in which they have the same colour. Hence $P(N, \lambda) = P(T, \lambda) + P(\phi_A(T), \lambda)$, and similarly $P(N, \lambda) = P(\theta_A(T), \lambda) + P(\psi_A(T), \lambda)$.

It is sufficient, following Birkhoff and Lewis, to specialize I and II to the cases in which there is only one vertex inside the separating digon or triangle. Equations (1) and (2) then reduce to

$$P(T, \lambda) = (\lambda - 2)P(T_2, \lambda) \quad \text{and} \quad P(T, \lambda) = (\lambda - 3)P(T_1, \lambda),$$

respectively.

Rules I, II and III can be used to construct lists of chromials of triangulations. One such list is given in the Birkhoff-Lewis paper. It extends up to 17 vertices, though completeness is not claimed beyond 10. Triangulations with separating digons or triangles are omitted from such lists.

Other simplifications are customarily made. The chromial of a planar triangulation T always divides by $\lambda(\lambda - 1)(\lambda - 2)$, and except in the rare Eulerian case by $(\lambda - 3)$. The chromials are therefore divided by $\lambda(\lambda - 1)(\lambda - 2)(\lambda - 3)$. They are moreover expressed in terms of $u = \lambda - 3$, because this change is found to give smaller coefficients. The transformation converts $P(T, \lambda)$ into a polynomial $Q(T, u)$ in u, called the Q-chromial of T. Normally it is the Q-chromial that is tabulated.

Ruth Bari, in her thesis [1], lists the Q-chromials of all the planar triangulations, having no vertex of valency less than five, with 19 or fewer vertices.

At this stage we should note some important properties of chromials with very simple inductive proofs based on I, II and III. Let $\alpha_0(T)$ be the number of vertices of a planar triangulation T. Then the coefficients in $P(T, \lambda)$ are integers. They are non-zero from the coefficient 1 (of $\lambda^{\alpha_0(T)}$) down to the coefficient of λ^1. The coefficients of the other powers of λ are zero. Moreover, the non-zero coefficients alternate in sign.

Birkhoff and Lewis developed the theory by trying to generalize the identities we have listed as I and II. Consider a planar triangulation S with a separating quadrilateral $Q = abcd$. Let D_1 be the map, with one quadrangular face, obtained from it by erasing the edges and vertices inside the quadrilateral. Similarly, let D_2 be the map, with a quadrangular outer face, obtained from S by erasing the edges and vertices outside Q. Ignoring the quadrangular face, we can regard each of the new maps as equivalent to a triangulated disc, bounded by the quadrilateral Q.

Birkhoff and Lewis introduced some polynomials, called constrained chromials, associated with D_1 and D_2. The first constrained chromial, $A_1(i)$ of D_i is the number of λ-colourings of D_i for which the four vertices of Q have all different colours. The second, $A_2(i)$ is the number of λ-colourings of D_i in which a and c have the same colour while b and d have different colours. $A_3(i)$ is the number in which b and d have the same colour while a and c have different colours. $A_4(i)$ is the number in which only two colours are used in Q.

It is easy to express $P(S, \lambda)$ in terms of the constrained chromials of D_1 and D_2:

$$P(S, \lambda) = \frac{A_1(1)A_1(2)}{\lambda(\lambda - 1)(\lambda - 2)(\lambda - 3)} + \frac{(A_2(1)A_2(2) + A_3(1)A_3(2))}{\lambda(\lambda - 1)(\lambda - 2)}$$

$$+ \frac{A_4(1)A_4(2)}{\lambda(\lambda - 1)} . \tag{4}$$

It is desirable however to express the constrained chromials in terms of free chromials, that is ordinary chromials of ordinary triangulations simply derived from D_1 and D_2.

In the case of D_1, we can obtain a triangulation T by subdividing the quadrilateral into two triangular faces by a diagonal $A = ac$. We can use the notation of III to denote other triangulations associated with D_1 by $\theta_A(T)$, $\phi_A(T)$, and $\psi_A(T)$. Birkhoff and Lewis obtained the following identities. In them, we write $A_j(1)$ simply as A_j.

$$(\lambda^2 - 3\lambda + 1)A_1 = \lambda(\lambda - 3)P(T, \lambda) + (\lambda - 3)P(\phi_A(T), \lambda)$$

$$- (\lambda - 3)(\lambda - 1)P(\psi_A(T), \lambda), \tag{5}$$

$$(\lambda^2 - 3\lambda + 1)A_2 = P(T, \lambda) + (\lambda - 2)^2 P(\phi_A(T), \lambda)$$

$$- (\lambda - 2)P(\psi_A(T), \lambda), \tag{6}$$

$$(\lambda^2 - 3\lambda + 1)A_3 = P(T, \lambda) - (\lambda - 3)P(\phi_A(T), \lambda)$$

$$+ (\lambda - 3)(\lambda - 1)P(\psi_A(T), \lambda), \tag{7}$$

$$(\lambda^2 - 3\lambda + 1)A_4 = - P(T, \lambda) + (\lambda - 3)P(\phi_A(T), \lambda)$$

$$+ (\lambda - 2)P(\psi_A(T), \lambda). \tag{8}$$

There are analogous identities for D_2.

It is now possible to express $P(S, \lambda)$ directly in terms of the chromials of triangulations simply related to D_1 and D_2.

Birkhoff and Lewis made a similar analysis for a separating pentagon. Again they obtained identities giving constrained chromials linearly in terms of free chromials. Again the coefficients were polynomials in λ, and again the coefficient of each constrained chromial was $(\lambda^2 - 3\lambda + 1)$.

The problem of the separating hexagon (or 6-ring) was solved by D. W. Hall and D. C. Lewis in a paper published in 1948. They obtained 41 identities giving the 15 constrained chromials. The coefficient of each constrained chromial was $(\lambda^2 - 3\lambda + 1)(\lambda^3 - 5\lambda^2 + 6\lambda - 1)$, multiplied by 6 or $6(\lambda - 2)$ [7].

Now what is the point of all this? We can, of course, argue that chromials are interesting in themselves and that they even have applications in the study of certain physical models [5]. But I am thinking here of a somewhat narrower point of view: why are chromials considered relevant to the study of the Four Colour Problem?

It can hardly be a simple matter of analogy between 4 and other integral values of λ; the colouring problem for triangulations is altogether too easy at those other integers. Birkhoff and Lewis wrote as follows. "It is also hoped that the theory of the chromatic polynomials may be developed to the point where advanced analytic function theory may be profitably applied."

The present writer thinks that this hope is the real justification for the study of chromials in connection with the Four Colour Problem. If so we should at least allow λ to take all real values. Then from theorems about $P(T, \lambda)$ for $\lambda \neq 4$ we might be able to infer something new about $P(T, 4)$ by continuity. Or we might allow λ to take all complex values and use analyticity. There is no difficulty in so extending the range of values of λ. We do indeed initially define $P(T, \lambda)$ in terms of positive integers. But once that function of λ is determined as a polynomial, it takes a definite value for each real or complex value of λ.

This line of thought leads to one clear conclusion. In our study of chromials we should ignore the point $\lambda = 4$ and collect as much information as we can about other values of λ. We should try to discover a pattern in the behaviour of chromials at these other values, hoping indeed that the pattern will ultimately guide us back in triumph to the point $\lambda = 4$, but giving no premature emphasis to this point, or indeed to integral points in general.

Not all writers on chromials seem to have accepted this conclusion. For example, some do not think it worth while to point out that $P(T, \lambda)$ is non-zero whenever λ is a negative real number, though this is an immediate consequence of the rule of alternating sign for coefficients. In the Birkhoff-Lewis paper, on the other hand, it is emphasized that the theorem that $P(T, \lambda) > 0$ when $\lambda \geqslant 5$ holds for all real numbers in this region and not merely for the integers.

Birkhoff and Lewis suggested a conjecture to replace the Four Colour Conjecture, and it is genuinely a conjecture about polynomials, not solely concerned with the case $\lambda = 4$. It concerns the polynomials $(\lambda - 3)^n$,

$$Q_n(\lambda) = \frac{P(T, \lambda)}{\lambda(\lambda - 1)(\lambda - 2)}$$

and $(\lambda - 2)^n$, where $n + 3$ is the number of vertices of the planar triangulation T. These are to be expressed as polynomials in $x = \lambda - c$, where c is an arbitrarily chosen real number not less than 4. The conjecture asserts that, for each T, each coefficient of a power of x in $Q_n(\lambda)$ is not less than the corresponding coefficient in $(\lambda - 3)^n$ and not greater than the corresponding coefficient in $(\lambda - 2)^n$. This conjecture implies the Four Colour Conjecture but, as far as is known, is not implied by it. This "Birkhoff-Lewis Conjecture" has undoubtedly stimulated much research on chromials. Ruth Bari stated as the main result of her thesis that she had proved the Birkhoff-Lewis Conjecture for all trivalent maps of fewer than 20 faces.

The present writer has to acknowledge that he does not feel much enthusiasm for conjectures even stronger than the Four-Colour one. Admittedly, some propositions become easier to prove by induction when they are strengthened in the proper way. But he pessimistically imagines that we must do much more hard work deepening the theory of chromials for general λ before we shall be in a position to bring off any such coup with the Four Colour Problem.

It is suggested then that the correct strategy is to study all values of λ except 4, and even to ignore the point $\lambda = 4$. That is a strong point, and we should attack the theory of chromials at its weak points, when and if we discover them. As for the problem of finding weak points, one obvious suggestion is that we should look for them among the zeros of the chromials. In advocating this procedure we are following D. W. Hall, J. W. Siry and B. R. Vanderslice. These authors published a paper in 1965 giving the Q-chromial, in terms of face-colourings, of the truncated icosahedron, [8]. In this paper there is a table of all the 28 zeros, 4 real and 24 complex.

An investigation of the zeros of chromials of planar triangulations was later carried out at Waterloo, using first the chromials tabulated by Ruth Bari, and then some generously made available by D. W. Hall. The latter represented intermediate stages of the work on the truncated icosahedron. The computer results showed one outstanding regularity. Each of the chromials studied had a zero close to $(3 + \sqrt{5})/2$, that is $\tau + 1$ where τ is the golden ratio. Reference to the Hall-Siry-Vanderslice paper showed that one of its four real zeros agreed with $\tau + 1$ to eight places of decimals. There seemed also a tendency for zeros to occur near $\lambda = 3.247$. It became customary at Waterloo to refer to $\tau + 1$ as the "golden root," and to the hypothetical real number to which zeros around 3.247 were approximating as the "silver root." These results were reported by G. Berman and W. T. Tutte in 1969, [2].

A letter to D. W. Hall brought the reply that he had not noticed that his table of roots was auriferous, but that he was now in a position to identify the silver root. He observed that the polynomial $\lambda^2 - 3\lambda + 1$, occurring as the coefficient of each constrained chromial in the Birkhoff-Lewis equations for the 4-ring, ((5)–(8)), had the golden root as one of its zeros. What more natural then that we should look for the silver root among the zeros of $\lambda^3 - 5\lambda^2 + 6\lambda - 1$, a factor of the corresponding coefficients in the equations of the 6-ring? Indeed this polynomial has a zero at $\lambda = 3.24698$, in close agreement with one of the Hall-Siry-Vanderslice real roots.

It seemed that two weak points had been detected. The weakness of the point $\lambda = \tau + 1$ was soon further demonstrated by the proof of several theorems, valid at the roots of $\lambda^2 - 3\lambda + 1 = 0$, but not valid for general λ. One of these is similar in form to (3).

$$P(T, \tau + 1) + P(\theta_A(T), \tau + 1)$$
$$= \tau^{-3}(P(\phi_A(T), \tau + 1) + P(\psi_A(T), \tau + 1)). \quad (9)$$

D. W. Hall pointed out that this theorem could be derived from the Birkhoff-Lewis equations for the 4-ring by putting $\lambda = \tau + 1$, and using (3). He modified the new inductive proof of (9) to simplify the derivation of the Birkhoff-Lewis equations for the 4-ring and 5-ring, [6].

Another theorem, proved in [11], is as follows:

$$|P(T, \tau + 1)| \leqslant \tau^{5-k}, \quad (10)$$

where k is the number of vertices of the triangulation T. This result may remain true, as regards absolute magnitude, but becomes much less impressive, when we replace $\tau + 1$ by the other zero $\tau^* + 1 = (-\tau^{-1} + 1)$ of $\lambda^2 - 3\lambda + 1$. For when k is large τ^{5-k} is very small but $|(\tau^*)^{5-k}|$ is very large. The theorem seems to provide an adequate explanation of the observed tendency of chromials of triangulations to have zeros near $\tau + 1$. In this connection it is of interest that $P(T, \tau + 1)$ never takes the value zero. This result depends on a theorem, proved by a simple induction, that the chromial of a connected loopless graph must be non-zero everywhere in the open interval $0 < \lambda < 1$. So in particular $P(T, \tau^* + 1)$ is non-zero. It follows that $P(T, \tau + 1)$ is non-zero. In a similar way we find that $P(T, \lambda)$ is non-zero at the roots of $\lambda^3 - 5\lambda^2 + 6\lambda - 1 = 0$, including the silver root.

We state one more theorem on the golden root:

$$P(T, \tau + 2) = (\tau + 2)\tau^{3k-10}P^2(T, \tau + 1). \quad (11)$$

It is proved in [12]. Theorems (10) and (11) can be used to check tables of chromials. Theorem (11) suggests that $\lambda = \tau + 2$ may be another weak point in chromial theory. Since $P(T, \tau + 1)$ is non-zero it follows from (11) that $P(T, \tau + 2)$ is positive, for every planar triangulation T. It is natural to make the conjecture that $P(T, \lambda)$ is positive whenever $\lambda \geqslant \tau + 2$, but there are counter-examples. The weakness of the silver root is not so clearly established. By substituting it in the equations of the 6-ring we can obtain a linear relation between chromials which is valid for the silver root but not for general λ. It is analogous to (9), but more complicated. It is noteworthy that the 15 equations given by Hall and Lewis reduce to one when this substitution is made. This seems to have been first observed by A. M. Hobbs at Waterloo.

At this stage S. Beraha came forward with the suggestion that all the numbers

$$B(n) = 2 + 2 \cos(2\pi/n), \qquad (12)$$

where n is an integer greater than 1, considered as values of λ, were weak points. Certainly the numbers $B(2)$, $B(3)$ and $B(4)$, that is 0, 1 and 2, are points of special significance; at these values of λ the chromial is zero for every planar triangulation T. $B(5)$ is the golden root and $B(6)$ is 3.

The value 3 of λ has some remarkable properties that are perhaps not yet properly appreciated. We know of course that $P(T, 3) = 0$ unless T is Eulerian, having every vertex of even valency. If T is Eulerian, then $P(T, 3) = 6$. There is essentially only one 3-colouring, but we have to allow for the six permutations of the three colours. It is a curious fact that $\lambda = 3$ is often not merely a zero but a multiple zero of $P(T, \lambda)$. Examples leap to the eye from the Birkhoff-Lewis table of Q-chromials. S. Beraha tells of triangulations with amazingly high values of the multiplicity. He has found families of such triangulations in which the multiplicity increases linearly with the number of faces.

This property of the number 3 is the more remarkable in that Q-chromials seem only very rarely to factorize into simpler polynomials (with integral coefficients). Apart from factorizations involving $(\lambda - 3)$ only two cases are known to the present writer. Triangulations with separating digons or triangles are of course not to be counted. The two Q-chromials in question are among those calculated by Hall, Siry and Vanderslice. Each one divides by $u^2 - u + 1$. It seems that we should add the complex number $3 - \omega$ to our list of weak points. Perhaps in this rarity of factors we encounter a really deep property of chromials, of which the apparent absence of the factor $(\lambda - 4)$ is only a special case.

Continuing with the numbers of the Beraha sequence we observe that $B(7)$ is the silver root. $B(8)$ is $2 + \sqrt{2}$, which is fairly close to the real root $\lambda = 3.41539930$ found by Hall, Siry and Vanderslice for the truncated icosahedron. $B(9)$ is $2 + 2 \cos 40^0$, that is 3.532. With a due allowance for diminishing accuracy this is perhaps not too far from the remaining real root 3.52004593 of the truncated icosahedron. $B(10)$ is $\tau + 2$, a value of λ that we have already encountered in (11).

These considerations have convinced the present writer that the significance of the Beraha sequence should be admitted, as a working hypothesis. This admission leads to the following prediction:

P. *For each integer $n \geqslant 2$ the n-ring is associated with a linear relation between chromials, the "free chromials" of Birkhoff and Lewis, which is valid for $\lambda = B(n + 1)$ but is not valid for all λ.*

Such a linear relation should be valid for all the roots of the minimal equation of $B(n + 1)$; Beraha is actually offering us a set of weak points that is dense in the interval $0 < \lambda < 4$. The prediction is trivially satisfied for $n = 2$ and $n = 3$, the linear relation being then $P(T, \lambda) = 0$. It is of course satisfied also for $n = 4$ and $n = 6$.

It is not difficult to verify the prediction in the case $n = 5$, $B(n + 1) = 3$. Consider a planar map N with a pentagonal face $a_1 a_2 a_3 a_4 a_5$. We suppose all the other faces, including the outer one, to be triangles. We regard the suffices as residues mod 5. We define Z_i as the triangulation obtained from N by taking the diagonals $a_i a_{i+2}$ and $a_i a_{i+3}$ as new edges. (See Figure 4.) We define Y_i as the triangulation obtained by identifying a_{i-1} and a_{i+1} and then deleting the original edge $a_i a_{i-1}$.

Now Z_1 is Eulerian if and only if a_3 and a_4 are the only vertices of odd valency in N. But Y_3 is Eulerian if and only if either a_2 and

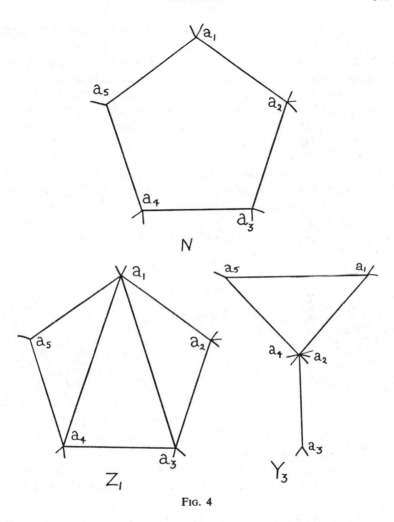

FIG. 4

a_3, or a_3 and a_4, are the only odd vertices in N. From these and the symmetrically related propositions we deduce that

$$P(Y_3, 3) = P(Z_1, 3) + P(Z_5, 3). \qquad (13)$$

This is the required linear identity. D. W. Hall has pointed out to the writer that it can also be derived from the Birkhoff-Lewis equations for the 5-ring.

In a recent paper [9], D. W. Hall gives a partial solution of the problem of the 7-ring. He says, "A new Beraha number appears exactly where it should." S. Beraha points out that his sequence converges to the number 4. If we can construct a theory of these allegedly weak points we can even hope to infer properties of $P(T, 4)$ by taking limits as $n \to \infty$.

We are thus able to conclude this account on a note of optimism. Let us continue the work done at the gold and silver roots, following the Beraha sequence further and further as it advances along the real axis.

At the end will be the Four Colour Problem.

REFERENCES

1. Bari, Ruth A., *Absolute reducibility of maps of at most 19 regions.* Thesis, Johns Hopkins, 1966.

2. Berman, G., and W. T. Tutte, "The golden root of a chromatic polynomial," *J. Combinatorial Theory*, 6 (1969), 301–302.

3. Birkhoff, G. D., "A determinant formula for the number of ways of coloring a map," *Ann. of Math.*, 14 (1912), 42–46.

4. Birkhoff, G. D., and D. C. Lewis, "Chromatic polynomials," *Trans. Amer. Math. Soc.*, 60 (1946), 355–451.

5. Fortuin, C. M., and P. W. Kasteleyn, "On the random-cluster model," (to be published).

6. Hall, D. W., "On golden identities for constrained chromials," *J. Combinatorial Theory*, 11 (1971), 287–298.

7. Hall, D. W., and D. C. Lewis, "Coloring six-rings," *Amer. Math. Soc. Transl.*, 64 (1948), 184–191.

8. Hall, D. W., J. W. Siry, and B. R. Vanderslice, "The chromatic polynomial of the truncated icosahedron," *Proc. Amer. Math. Soc.*, 16 (1965), 620–628.

9. Hall, D. W., "Coloring Seven-Circuits," Graphs and Combinatorics, New York: Springer-Verlag vol. 406, 1974, 273–290.

10. Ore, O., *The four-color problem*, New York: Academic Press, 1967.

11. Tutte, W. T., "On chromatic polynomials and the golden ratio," *J. Combinatorial Theory*, **9** (1970), 289–296.

12. ——, "The golden ratio in the theory of chromatic polynomials," *Annals of the New York Academy of Sciences*, **175**, 1 (1970), 391–402.

13. Whitney, H., "The coloring of graphs," *Ann. of Math.*, **33** (1932), 688–718.

KEMPE CHAINS AND THE FOUR COLOUR PROBLEM*

Hassler Whitney and W. T. Tutte

1. INTRODUCTION

In October 1971 the combinatorial world was swept by the rumour that the notorious Four Colour Problem had at last been solved, —that with the help of a computer it had been demonstrated that any map in the plane can be coloured with at most four colours so that no two countries with a common boundary line are given the same colour.

The first ostensible proof of the conjecture to be published was that of A. B. Kempe [4]. This appeared in 1879 and was accepted for a decade. A flaw in the argument was pointed out in 1890 by P. J. Heawood [2], whose papers on the problem span the next sixty years. After Heawood's first paper, mathematicians began to suspect that the Four Colour Problem was of surpassing difficulty; perhaps it was to be ranked with Fermat's Last Theorem and the

*This work was partly supported by a grant from the National Research Council of Canada.

Riemann Hypothesis. G. D. Birkhoff once told one of the authors that every great mathematician had at some time attempted the Four Colour Conjecture, and had for a while believed himself successful. "Proofs" of the Conjecture are still written every now and then, and occasionally one gets published. Anyone now having a proof that he wishes to be taken seriously would be well advised to write it out clearly and in full logical detail, so that any mathematician willing to spend enough time on it will be able to check it.

For the history and present status of the Four Colour Problem reference may be made to the book by O. Ore [5] and the recent article by T. L. Saaty [7].

The rumour mentioned above arose from the work of Y. Shimamoto, who claimed a proof based on the work of H. Heesch [3]. Heesch has for years been studying the "reducibility" of maps. He shows that some configurations of countries have a property that he calls "D-reducibility," and he has a method whereby a given configuration can be tested for this property. It requires much computer time. Shimamoto, on the assumption that the Four Colour Conjecture was false, showed that there must be a non-colourable map M containing a configuration H that had already passed the computer test for D-reducibility. He then arrived at a contradiction by showing that the D-reducibility of H implied the 4-colourability of M. This argument seemed to prove the Conjecture. The burden of proof was not now on a few pages of close reasoning, but on a computer!

This method of proof was greeted by the present authors (independently) first with some misgivings and then with real scepticism. It seemed to both of us that if the proof was valid it implied the existence of a much simpler proof (to be obtained by confining one's attention to one small part of M), and that this simpler proof would be so simple that its existence was incredible. The present paper is essentially the result of our attempts to give a proper mathematical form to our objection.

We found no essential flaw in Shimamoto's reasoning. (It was later set out "clearly and in full logical detail" in an article circulated by W. R. G. Haken.) We therefore decided that the computer result must be wrong. (We learned later that a repro-gramming of the computer had indeed given the result that H was not D-reducible.) However it turned out that the basic result of Shimamoto was simply that a certain plane graph G_{12} (see Figure 12) cannot be vertex-coloured in four colours so that the boundary of each pentagonal face uses all four colours. (The reader can check this at once; see the proof near the end of Section 9.)

Is there a moral to be drawn? Perhaps it is that if you are really interested in the results of a study you should give it an analysis in depth and try to understand fully its implications. In fact the strength of present-day mathematics owes much to this principle. It seems that in the present case deductions were made from the accepted D-reducibility of H without an adequate understanding of what D-reducibility is. With this understanding the deductions become incredible and are seen to constitute a proof by *reductio ad absurdum* that H is not in fact D-reducible.

In this paper we give a general description of this type of approach to the Four Colour Problem. We define Kempe chains, and we point out some things that can be done with Kempe chains and some things that cannot. The exposition is intended to be generally understandable, not requiring any special mathematical preparation.

2.   PRELIMINARIES.

First we note a dual formulation of the problem, easier to visualize and use. Consider any map M on the plane or the sphere. In each country we mark the capital; this will be a vertex of a graph G. If two countries have a common border we join their capitals by a railroad across it. This railroad is an edge of G. (The terms "vertex" and "edge" are taken from the theory of poly-hedra.) Colouring the map M is equivalent to colouring the vertices of G so that no two vertices joined by an edge are of the same colour. Any plane graph comes from a map in this way, provided that no edge is a "loop," i.e., joins a vertex to itself. From now on, all the plane graphs that we consider are to be assumed loopless. The Four Colour Conjecture can now be formulated as follows: any plane graph can be 4-coloured. This is the formulation used in Ore's book.

The edges of a connected plane graph cut the plane into regions that we call the *faces* of G; one of these is the *outside* face. If we project the plane stereographically onto the sphere the outside face is made to surround the North Pole. By choosing another diameter as the axis of the sphere we can arrange that any desired face contains the North Pole, so on projecting back into the plane we can arrange that any desired face becomes the outside face.

A *face-boundary* of G is the boundary of a face of G. A *circuit* is a graph defined by the set of edges and vertices of a simple closed curve. We note that a face-boundary is not necessarily a circuit; it may include an edge or vertex whose removal disconnects G. (An edge of this kind is called an *isthmus* of G.) If Q is a face-boundary of G we say for short that G is *face-bounded* by Q. If Q is a circuit the remainder of G lies on one side of Q, the inside or the outside, unless G consists solely of Q. A circuit Q made up of edges of G is *vertex-separating* in G if G has a vertex inside Q and a vertex outside Q. Similarly it is *edge-separating* if G has an edge inside Q and an edge outside Q.

A *triangulation* of the plane is a plane graph G whose faces are all triangles, that is whose face-boundaries consist of three edges each. The *valency* $v(A)$ of a vertex A is the number of edges having A as an end. Most commonly these edges go to distinct vertices, but it is not necessary to impose this as part of the definition. A *k-wheel* is a graph W formed from a circuit of k edges by adjoining a new vertex A and then joining A to each vertex of the circuit by a single new edge. A is the *hub* of the wheel, the circuit is the *rim* and the new edges are the *spokes*. In a triangulation in which at most one edge joins any two vertices (and which has at least 4 vertices) each vertex is the hub of a wheel.

A graph is *k-chromatic* if it can be coloured in k colours but not in fewer. A *full* k-colouring of a graph is one that uses exactly k colours. A 3-colouring is usually accepted as a special kind of 4-colouring, but it is not a full 4-colouring.

3. THE EULER FORMULA.

In the rest of this paper the symbol G is used to denote a non-null connected plane graph in which there is no loop. Let such a graph G have N_v vertices, N_e edges and N_f faces. Then the Euler formula is

$$N_v - N_e + N_f = 2. \tag{1}$$

The formula can be proved by induction on $N = N_v + N_e + N_f$. The smallest possible value of N is 2. This occurs only when G has a single vertex, no edge and a single face (which is an "agon"). In this simple case the Euler formula holds. In the general case suppose first that G has a circuit. We drop out one edge of the circuit. This decreases N_e and N_f by one each, and leaves N_v unchanged. By the inductive hypothesis (1) holds for the resulting graph, and therefore it holds for G. If G has no circuit we can find a vertex joined to just one other. Dropping out this vertex and its edge reduces N_v and N_e each by one and leaves N_f unchanged. (It is 1.) Again the formula follows for G.

We show next that for triangulations of the plane we have

$$N_f = 2N_v - 4, \qquad N_e = 3N_v - 6. \tag{2}$$

In each face draw a new face-boundary just inside the original one. We have drawn $3N_f$ new "edges" and two of these are beside each former edge. Hence $3N_f = 2N_e$. Combining this with (1) we obtain (2).

We can carry the argument further. Each vertex X is on $v(X)$ edges and each edge is on two vertices. Hence $\Sigma v(X) = 2N_e$. Let us write $v'(X) = v(X) - 6$. Then by (2) we have

$$\sum v'(X) = \sum v(X) - 6N_v = -12. \tag{3}$$

Hence each triangulation has vertices of valency less than 6. If each valency is at least 5 then there are at least 12 vertices of valency 5, and more according as vertices of valency greater than 6 are present.

4. KEMPE CHAINS.

Consider a fixed set $\{\alpha, \beta, \gamma, \delta\}$ of four colours. There are three *colour partitions* of this set into two pairs of colours, namely $(\alpha\beta, \gamma\delta)$, $(\alpha\gamma, \beta\delta)$ and $(\alpha\delta, \beta\gamma)$. Let the graph be 4-coloured, and let Γ denote its 4-colouring. For each pair of colours, say $\{\alpha, \beta\}$, let $G_{\alpha\beta}$ denote the subgraph of G consisting of the vertices coloured α or β and the edges joining them. Now the "components" of a graph are its maximal connected subgraphs. We refer to the components of $G_{\alpha\beta}$ as the *Kempe chains belonging to* the unordered pair $\{\alpha, \beta\}$ in Γ. We call them also the $\alpha\beta$-*chains* of Γ. It is possible for a Kempe chain to consist of a single vertex.

A *Kempe interchange* in Γ, with respect to the colour-partition $(\alpha\beta, \gamma\delta)$ is an interchange of the colours α and β in one of the $\alpha\beta$-chains, or of the colours γ and δ in one of the $\gamma\delta$-chains. If there are m Kempe chains belonging to $\{\alpha, \beta\}$ or $\{\gamma, \delta\}$ then we obtain 2^m colourings of G by performing or not performing the interchange in each of the m chains. Each interchange leaves the $\alpha\beta$-chains and the $\gamma\delta$-chains unaltered.

We note however than an interchange in an $\alpha\beta$-chain of Γ must alter the system of $\alpha\gamma$-chains for example. Some of them are destroyed or new ones appear, or both. We proceed to state an obvious but important theorem about Kempe chains.

THEOREM 4.1: *Let a plane graph U be face-bounded by a circuit Q. Let the vertices A, B, C, D lie in that order in Q. Let U be 4-coloured. Then if there is an $\alpha\beta$-chain of the 4-colouring containing A and C but not B or D there can be no $\alpha\beta$-chain or $\gamma\delta$-chain of the 4-colouring containing both B and D. (See Figure 1.)*

Theorem 4.1 has the following elementary consequence:

THEOREM 4.2: *Let a plane graph U be face-bounded by a quadrilateral $Q = ABCD$, and let Γ be a 4-colouring of U using all four colours in Q. Then we can find a Kempe interchange in Γ that transforms Γ into a 4-colouring of U using exactly three colours in Q.*

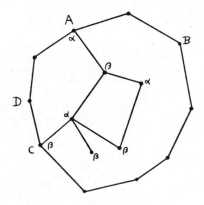

FIG. 1

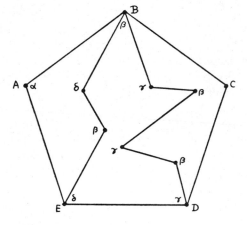

FIG. 2

Proof: Let A, B, C and D have colours α, β, γ and δ respectively. If the $\alpha\gamma$-chain containing A does not contain C, then a Kempe interchange in it has the required effect. In the remaining case the $\beta\delta$-chain containing B does not contain D, by 4.1, and we can use the Kempe interchange in it.

5. ELEMENTARY REDUCTIONS.

If the Four Colour Conjecture is false, there must be a least integer N such that a 5-chromatic plane graph of N vertices exists, and no two distinct edges of this graph have the same pair of ends. Such a graph is called *minimal*. T. L. Saaty calls N the *Birkhoff Number*, and completes his definition by saying that $N = \infty$ if the Conjecture is true. If the Four Colour Conjecture is assumed for all finite plane graphs, it follows at once for infinite ones. This is an easy consequence of the Cantor "diagonal process." In this Section we assume that N is finite, and we study the properties of minimal graphs.

Obviously $N \geqslant 4$. In 1938 C. E. Winn showed that $N \geqslant 36$ [9]. In a paper published in 1970, O. Ore and J. Stemple claim to have shown that $N \geqslant 40$ [6]. Their numerical calculations are too lengthy for publication in a Journal, but are available in the library of the Mathematics Department at Yale University.

Besides the minimal graph, investigators of colouring problems often use the concept of a *critical graph*. For our purposes we can define a critical graph as a 5-chromatic plane graph whose proper subgraphs are all 4-colourable. A minimal graph, we shall show, is critical, but we are not entitled to assert that any critical graph must be minimal.

Most studies of the 4-colour problem have been studies of what minimal graphs must be like, with the ultimate object of showing that no such graphs exist. We give some sample properties.

THEOREM 5.1: *In a minimal graph G each face-boundary is a circuit.*

Proof: Suppose the boundary P of some face F is not a circuit. Let us go around it close to the boundary; we pass near some vertex X at least twice. We can go from X through F and back to X in such a way as to traverse a simple closed curve having vertices of G both inside and outside. Thus G is the union of two plane graphs G_1 and G_2 having only the vertex X in common. Each of these can be 4-coloured since each has fewer than N vertices. Having 4-coloured G_1 and G_2 we can permute the colours in G_1 so as to make the two 4-colourings agree at X. We can then combine the two 4-colourings into a 4-colouring of G. This contradiction establishes the theorem.

THEOREM 5.2. *A minimal graph G is a triangulation.*

Proof: Assume the contrary. Then G has a face, which may be taken as an inside face with a boundary circuit Q (see 5.1) containing four vertices A, B, C and D, in that order. Now we cannot have edges AC and BD outside Q. We may therefore assume without loss of generality that no edge joins A and C. We may therefore pull A and C together inside Q and let them become a single vertex, so obtaining a plane graph G'. This has fewer than N vertices and is thus 4-colourable. But a 4-colouring of G' obviously determines one of G, and we have a contradiction.

We can argue from 5.2 that it is sufficient to prove the Four Colour Conjecture for triangulations. If it is true for them, it must be true for all plane graphs. Accordingly most papers on the subject are concerned with triangulations only (or their duals, if face-colourings are being considered).

THEOREM 5.3: *A minimal graph has no vertex of valency less than 5.*

Proof: Suppose a minimal graph G to have a vertex V of valency < 5. Let G' be the plane graph obtained from it by deleting V and its incident edges. Since G' has fewer than N vertices it can be 4-coloured. If V is joined to at most 3 other vertices in G we can obviously extend any 4-colouring of G' to a 4-colouring of G. In the remaining case G' is bounded by a quadrilateral $ABCD$, by 5.2. By 4.2 we can find a 4-colouring of G' that uses only three colours in $ABCD$, and this can be extended as a 4-colouring of G, —a contradiction.

THEOREM 5.4: *A minimal graph G has no vertex-separating circuit of fewer than 5 vertices.*

Proof: Suppose G has such a vertex-separating circuit Q. Let Q bound the subgraphs G_1 and G_2, whose union is G and whose intersection is Q. We can find 4-colourings Γ_1 and Γ_2 of G_1 and G_2 respectively, since each of these graphs has fewer than N vertices. If Q has at most three vertices we can permute the colours in Γ_1 so as to make Γ_1 and Γ_2 agree in Q. We can then combine Γ_1 and Γ_2 to make a 4-colouring of G.

From now on we may suppose Q to be a quadrilateral $ABCD$. If all four colours are used in Q in both Γ_1 and Γ_2 we can permute and combine, much as before, to obtain a 4-colouring of G. We may therefore assume, without loss of generality, that there is no 4-colouring of G_1 for which all four colours appear in Q.

Let us modify G_2 by joining A and C across the face Q. The resulting graph has fewer than N vertices and is thus 4-colourable. We deduce that G_2 has a 4-colouring with at least 3 colours in Q. Hence we can choose Γ_2 to have exactly 3 colours in Q, by 4.2. Without loss of generality we can suppose B and D to have the same colour α in Γ_2, while A and C have distinct colours.

Considering the effect of joining A and C across Q in G_1 we find that we can choose Γ_1 so that A and C have distinct colours. Then B and D must have the same colour, by the restriction we have been able to impose on G_1. We can now permute colours in Γ_1 so as to make it agree with Γ_2 in Q, and then combine Γ_1 and Γ_2 into a 4-colouring of G.

In every case we have found a contradiction.

By a similar but more complex argument G. D. Birkhoff showed that a minimal graph has no circuit of 5 edges separating at least two vertices from at least two others [1].

There is a similar theory of 5-colourings in which one proves, in the manner of 4.2 and 5.3, that a 5-minimal graph has no vertex of valency less than 6. It follows that 5-minimal graphs do not exist, by § 3; the "Five Colour Theorem" is true.

THEOREM 5.5: *A minimal graph G is critical.*

Proof: Let A be any edge of G. Form G' from G by deleting A. By our definition of a minimal graph there is no second edge of G joining the ends of A. Accordingly the edge A of G can be contracted to a single vertex so as to transform G into a new connected loopless plane graph G'' with one vertex fewer. But G'' has a 4-colouring, by the minimality of G, and this evidently determines a 4-colouring of G'. Hence each subgraph of G' is 4-colourable. Since A was chosen arbitrarily it follows that every proper subgraph of G is 4-colourable, that is, G is critical.

6. K-REDUCIBILITY.

Consider an edge-separating circuit Q (as defined in Section 2) in a plane graph G. It decomposes G into two plane graphs U and V, each having Q as a face-boundary. Each of U and V has one new face bounded by Q, and its other faces are faces of G. The graphs U and V have G as their union and Q as their intersection. Of course each of U and V has an edge not in Q.

Many studies of the Four Colour Problem are concerned with such figures, which we call Q-decompositions of G. Typically it is shown that if V has a specified structure, then G cannot be minimal, whatever the structure of U. We express this property of the given V, with its boundary Q, by saying "(V, Q) is *reducible*." As far as we know at present it is not inconsistent with V being 5-chromatic, but of course this possibility has not been realized in practice.

Let U be a plane graph bounded by a circuit Q. Let Γ be a 4-colouring of Q. It may happen that there is a 4-colouring Γ' of U that reduces to Γ on Q. If so we say that Γ is *U-extensible*, and that Γ' is a *U-extension* of Γ.

Now let Γ be a 4-colouring of Q, and let S be a set of 4-colourings of Q. It may happen that Γ is U-extensible and that every U-extension of Γ can be transformed, by a succession of Kempe interchanges in U, into a U-extension of some member of S. If so we say that Γ is *U-immersible* in S.

Let Γ be a 4-colouring of a circuit Q, and let S be a set of 4-colourings of Q. It may happen that Γ is U-immersible in S for every plane graph U bounded by Q and such that Γ is U-extensible. If so we say simply that Γ is *immersible* in S.

Let V be a plane graph face-bounded by a circuit Q and having at least one edge not in Q. Let S be the set of all V-extensible 4-colourings of Q. Evidently each member of S is immersible in S. We say that V is *K-reducible* with respect to Q, or, briefly, that (V, Q) is *K*-reducible, if every 4-colouring of Q is immersible in S.

In applications of the last definition we usually think of Q as bounding an inner face of V. But of course the distinction between "inner" and "outer" is only a matter of convenience.

THEOREM 6.1: *If (V, Q) is K-reducible, then (V, Q) is reducible.*

Proof: Let V be represented as part of a Q-decomposition of a plane graph G. Then Q is edge-separating in G. Assume that G is minimal. Then, by 5.5, U has a 4-colouring Γ', a U-extension of a 4-colouring Γ of Q. Since (V, Q) is K-reducible Γ is transformable into some 4-colouring Γ_1 of Q that is V-extensible, by a succession of Kempe interchanges in U starting with Γ'. Now Γ_1 has both a U-extension and a V-extension. Combining these we obtain a 4-colouring of G, which is a contradiction. Thus G cannot be minimal, for any U, and so (V, Q) is reducible.

If the 5-wheel could be proved reducible with respect to its rim, then the Four Colour Conjecture would be verified. For since no vertex of a minimal graph G can have valency less than 5, by 5.3, and since no vertex-separating circuit of G has fewer than 5 edges, by 5.4, it follows from § 3 that G has a vertex of valency 5 and that this is the hub of a 5-wheel contained in G.

Let us consider an attack on the K-reducibility of a 5-wheel, with rim $Q = ABCDE$. In Figure 2 we show part of a coloured plane graph U inside Q. Actually we try to build up a graph U so that the 4-colouring of Q shown in the figure is U-extensible, but is not U-immersible in the set S of all 4-colourings of Q using only 3 distinct colours. If we can carry out the proposed construction, then the 5-wheel is not K-reducible. If we can prove the construction impossible, K-reducibility is established and the Four Colour Conjecture is verified. Actually we shall achieve neither of these results. But let us assume that we have found a plane graph U of the required kind, and that we have a U-extension Γ of the 4-colouring of Q shown in the diagram. We note that all full 4-colourings of Q are equivalent to within rotations and reflections of the pentagon and permutations of the four colours.

First, if the $\beta\delta$-chain containing B does not contain E, then an interchange of β and δ in this chain removes β from Q. Hence we must construct U and Γ so that this chain also contains E. Similarly we must arrange that the $\beta\gamma$-chain containing B contains also D. (See Figure 2.) Because of the $\beta\delta$-chain, the $\alpha\gamma$-chain containing A does not contain D; interchange α and γ in it. Because of the $\beta\gamma$-chain the $\alpha\delta$-chain containing C does not contain E; interchange α and δ in it. At first sight it seems that we thus remove α from Q; the construction has been proved impossible and the Four Colour Conjecture follows.

This in fact was Kempe's proof. What is wrong? Simply that we may not be able to make both the interchanges called for. This point is clarified by Figure 3. In this figure the $\alpha\gamma$-chain from A contains a vertex of the $\beta\gamma$-chain from B to D. The interchange of α and γ breaks the latter chain and sets up a new $\alpha\delta$-chain from C to E. This makes it impossible to remove α from Q by an interchange of α and δ.

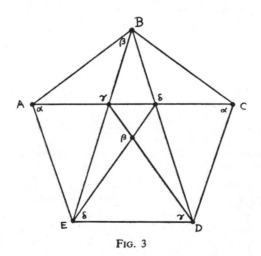

Fig. 3

In the graph shown we can remove a colour from Q by making first the suggested interchange of α and γ, and then operating on a $\beta\gamma$-chain and a $\gamma\alpha$-chain as shown in Figure 4. We ought now to complicate U so as to make this sequence of operations impossible. If the reader tries to do this we feel safe in saying that other ways of eliminating a colour from Q will appear, calling for further complication. If there is an end to this process it is not yet in sight.

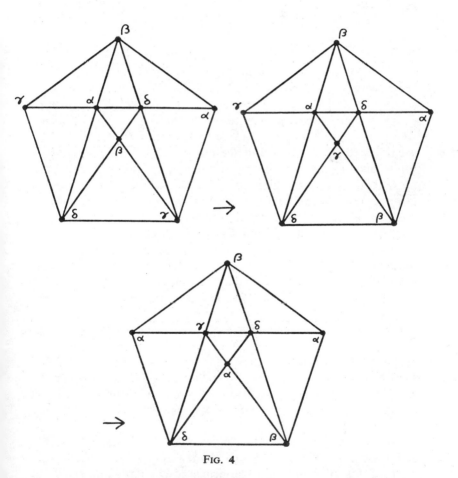

Fig. 4

7. D-REDUCIBILITY.

The difficulty in studying K-reducibility comes from the fact that if we make a Kempe interchange, Kempe chains from other colour partitions are altered. This leads to theoretical complications that mathematicians have so far been unable to resolve.

One of the present authors sought to make a modest first step toward resolving them through a theory of "parity," whereby the 4-colourings of a triangulation can be classified as "even" or "odd." He was able to show that the parity of a 4-colouring is invariant under Kempe interchanges [8]. But this result is not strong enough to be of much help in the theory of reducibility.

We now discuss a special kind of reducibility, called *D-reducibility*, introduced by H. Heesch. Its theory avoids any consideration of the effect of a Kempe interchange on the Kempe chains of other colour partitions. Heesch gives an algorithm whereby a given configuration can be tested for D-reducibility. As a hand method it is long and tedious, but it can be programmed for a computer. Heesch advocates the construction of a catalogue of D-reducible, and otherwise reducible, configurations, hoping that ultimately it can be shown that every triangulation contains a member of the list. If so it will be proved that no triangulation is minimal, and so that the Four Colour Conjecture is true.

We define a *near-triangulation* as a plane graph N in which at most one face is non-triangular. When we speak of a near-triangulation as being bounded or face-bounded by a circuit Q it is to be understood that if there is a non-triangular face it is the one bounded by Q.

Let Γ be a 4-colouring and S a set of 4-colourings of a circuit Q. It may happen that there is a colour partition Π with the following property: if U is any near-triangulation bounded by Q and if Γ is U-extensible, then there is a U-extension Γ' of Γ that can be transformed into a U-extension of some member of S by a succession of Kempe interchanges with respect to Π. If so we say that Γ is *simply immersible* in S.

We denote the set of all 4-colourings of Q that are simply immersible in S by $f(S)$. Evidently $S \subseteq f(S)$. Then $f^2(S) = f(f(S))$ is the set of all 4-colourings of Q simply immersible in

$f(S)$, and so on. A 4-colouring Γ is said to be *crudely immersible* in S if it belongs to $f^k(S)$ for some nonnegative integer k. ($f^0(S)$ $= S$). Its crude immersion in S is effected by a succession of simple immersions in sets $f^j(S)$ with decreasing j, and it is not necessary that these simple immersions shall all be associated with the same colour partition Π. We say that the set S is *dominant* if every 4-colouring of Q is crudely immersible in S.

Can one classify the dominant sets for any given circuit Q? That would seem to be an interesting question. Heesch does not solve it, but his algorithm will determine whether a given set of 4-colourings is dominant.

Let V be a near-triangulation bounded by a circuit Q and having at least one edge not in Q. We say that V is *D-reducible* with respect to Q, or, briefly, that (V, Q) is D-reducible, if the set of all V-extensible 4-colourings of Q is dominant.

THEOREM 7.1: *If (V, Q) is D-reducible, then (V, Q) is reducible.*

Proof: Adjoin to V any other near-triangulation U bounded by Q so as to form a triangulation G with Q as an edge-separating circuit. Assume that G is minimal. Then U has a 4-colouring Γ', a U-extension of a 4-colouring Γ of Q, by 5.5. Let S be the set of all V-extensible 4-colourings of Q. Then $\Gamma \in f^k(S)$ for some nonnegative integer k, since (V, Q) is D-reducible. Choose Γ', Γ and k so that k has the least possible value.

If $k > 0$ then, by the definition of $f^k(S)$, some U-extension of Γ can be transformed by a succession of Kempe interchanges, all with respect to the same colour partition, into a U-extension of some member Γ_1 of $f^{k-1}(S)$. This being contrary to the choice of k we deduce that in fact $k = 0$. This means that Γ has both a U-extension and a V-extension. Combining these extensions we obtain a 4-colouring of G.

From this contradiction we deduce that G is in fact not minimal. Thus (V, Q) is reducible.

As an example let V be a 4-wheel and Q its rim. Then S, the set of V-extensible 4-colourings of Q, consists of the non-full 4-

colourings of Q. By 4.2 each of the remaining 4-colourings of Q is in $f(S)$. Thus S is dominant and (V, Q) is D-reducible. On the other hand we have the following theorem:

THEOREM 7.2: *Let $Q = ABCDE$ be a pentagon, and let S be the set of all 4-colourings of Q using only 3 distinct colours. Then S is not dominant.*

Proof: Let Γ be any full 4-colouring of Q. Adjusting the notation we can represent it by Figure 5.

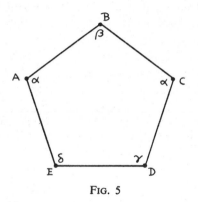

FIG. 5

Consider the three colour partitions in turn. In the case of $(\alpha\beta, \gamma\delta)$ whatever U we take, bounded by Q, the vertices A, B and C are in one $\alpha\beta$-chain and the vertices D and E are in one $\delta\gamma$-chain. Interchanges in this partition yield only full 4-colourings of Q. In the case of $(\alpha\gamma, \beta\delta)$ let us form U by adding an edge BD and an edge BE. Any Kempe interchange leaves both colours β and δ on BE, and both colours α and γ on CD. Again, only full 4-colourings of Q can be derived. By the symmetry of the figure, the same proof applies to the case $(\alpha\delta, \beta\gamma)$. We conclude that Γ is not in $f(S)$. In fact $f(S)$ must be identical with S, and so S is not dominant.

COROLLARY: *The 5-wheel is not D-reducible.*

THEOREM 7.3: *Let U be a near-triangulation bounded by a pentagon Q, and let U be 4-colourable. Suppose that no 4-colouring of Q using only three distinct colours is U-extensible. Then every full 4-colouring of Q is U-extensible.*

We prove this by "crude chaining", that is we do not need to consider the effect of our Kempe interchanges on the Kempe chains of other colour partitions.

Proof: We may assume that the full 4-colouring of Q shown in Figure 5 is U-extensible. There is a $\beta\delta$-chain in U from B to E, since otherwise an interchange would eliminate β from Q. An interchange on one $\alpha\gamma$-chain interchanges the colours of C and D, leaving A coloured α. We have thus moved the pair of vertices distinguished by a common colour two steps counter-clockwise round the pentagon. Repetition of this process gives five distinct full 4-colourings of Q, all U-extensible, from which all the full 4-colourings of Q can be derived by permutations of the four colours. The theorem follows.

8.   CHROMODENDRA.

The definitions of K-reducibility and D-reducibility suffer from one disadvantage. They require the consideration of an infinity of possible near-triangulations U bounded by Q. In the case of K-reducibility we may hope to overcome this disadvantage by proving general theorems about the interrelations of Kempe chains of colourings of U, but so far no suitable general theorems have been established. In the case of D-reducibility we shall show, following Heesch, that the possible near-triangulations U can be classified under a finite number of cases, and that each case can be dealt with by a finite argument. The question of whether a given V, bounded by Q, is D-reducible can thus be reduced to a finite problem.

Let Q be a circuit of n edges, and let Γ be a 4-colouring of Q. Of course Γ has its Kempe chains, these being subgraphs of Q. If α and β are any two of the four colours we define an $\alpha\beta$-*cluster* of Γ as a non-null set of Kempe chains of Γ belonging to the colour-pair $\{\alpha, \beta\}$. A *clustering* of Γ, with respect to the colour-partition $(\alpha\beta, \gamma\delta)$, is a family of $\alpha\beta$-clusters and $\gamma\delta$-clusters of Γ such that each Kempe chain of Γ belonging to $\{\alpha, \beta\}$ or $\{\gamma, \delta\}$ is contained in exactly one of them. A vertex of G is said to be *included* in a cluster C if it is a vertex of one of the Kempe chains making up C.

A clustering Z of Γ, with respect to $(\alpha\beta, \gamma\delta)$, is said to be *admissible* if it has the following properties:

(i) Let C_1 and C_2 be any two distinct members of Z. Let a_1 and b_1 be any two vertices included in C_1, and let a_2 and b_2 be any two vertices included in C_2. Then a_1 and b_1 do not separate a_2 from b_2 in Q.

(ii) Let C_1 be any cluster in Z and let L be any arc in Q such that no vertex of L is included in C_1 but each end of L is a vertex adjacent to a vertex that is included in C_1. Then there is a cluster C_2 in Z such that each end of L is included in C_2.

Two clusters C_1 and C_2 in a clustering Z are said to be *adjacent* in Z if there is an edge A of Q for which one end is included in C_1 and one in C_2. We then say that A is a *joining edge* of the two clusters. Evidently adjacent clusters are associated with complementary colour-pairs.

THEOREM 8.1: *Let C_1 and C_2 be adjacent clusters in an admissible clustering Z. Then C_1 and C_2 have exactly two joining edges A and B. Moreover the deletion of A and B decomposes Q into two disjoint connected graphs L_1 and L_2 such that L_1 contains all the vertices of Q included in C_1, and L_2 all those included in C_2.*

Proof: C_1 and C_2 have one joining edge A. Let its ends included in C_1 and C_2 be a_1 and a_2 respectively. Follow along Q from a_2, away from a_1, until we reach a last vertex b_2 not in C_1; the next edge B has its other end b_1 in C_1. This gives L_1 and L_2. (L_1 is an arc, unless $a_1 = b_1$.) By (ii), b_2 is in C_2. By construction L_2 has no vertices in C_1, and by (i) L_1 has no vertices in C_2. Thus (8.1) follows.

It is convenient to represent an admissible clustering Z of Γ, with respect to $(\alpha\beta, \gamma\delta)$ by a graph χ called its *chromodendron*. The vertices of χ represent the member-clusters of Z, and two vertices are joined by an edge if and only if the two member-clusters corresponding are adjacent. Two vertices are not to be joined by more than one edge.

If two vertices are adjacent in Q they are included in the same cluster, or in adjacent clusters, of Z. From this observation we deduce that χ is connected. From 8.1 we deduce that each edge of χ is an isthmus. Thus,

THEOREM 8.2: *Every chromodendron is a tree.*

The chromodendra of the admissible clusterings of Γ, with respect to $(\alpha\beta, \gamma\delta)$ will be called the chromodendra of Γ, with respect to $(\alpha\beta, \gamma\delta)$. If n is reasonably small there is no difficulty in making a list of all such structures for a given Γ and a given colour-partition.

Let Γ and Γ_1 be 4-colourings of the circuit Q and let χ be a chromodendron of Γ with respect to $(\alpha\beta, \gamma\delta)$. We say that Γ is *simply χ-transformable* into Γ_1 if it is transformed into Γ_1 by interchanging the two colours in each of the Kempe chains of Γ belonging to a single cluster, this cluster being represented by a vertex of χ. All the Kempe chains of this cluster are affected, but no other $\alpha\beta$-chains or $\gamma\delta$-chains. We note that after the operation χ remains as a chromodendron of Γ_1, with respect to $(\alpha\beta, \gamma\delta)$, and we can consider the application of a second simple χ-transformation to Γ_1. We say that Γ is *χ-transformable* into a 4-colouring Γ' of Q if Γ can be changed into Γ' by a succession of simple χ-transformations, all of course referring to the same chromodendron χ. We also express this by saying that χ *admits* a transformation of Γ into Γ'.

We proceed to relate the theory of chromodendra to that of
D-reducibility.

Let U be a near-triangulation bounded by a circuit Q. Let Γ be
a 4-colouring of Q and let Γ' be a U-extension of Γ. Let K be a
Kempe chain of Γ' belonging to the colour-pair $\{\alpha, \beta\}$. Suppose it
to have at least one vertex in common with Q. Then the intersection of K with Q is a union of one or more $\alpha\beta$-chains of Γ. The
$\alpha\beta$-chains of Γ contained in $K \cap Q$ thus constitute an $\alpha\beta$-cluster of
Γ. We call this the *residue* of K in Γ. Kempe chains of Γ' having no
vertex in common with Q are considered to have no residues in Γ.
Figure 6 shows Q bounding U, with the colourings Γ and Γ'. The
full lines represent the edges of the $\alpha\beta$-chains and $\gamma\delta$-chains of Γ'.
There are two $\alpha\beta$-chains. One of them has a residue consisting of
three $\alpha\beta$-chains of Γ. These have the vertex-sets $\{A\}$, $\{H\}$ and
$\{E, F\}$. The second $\alpha\beta$-chain of Γ' has a residue consisting of a
single Kempe chain of Γ, this having the vertex-set $\{C\}$. One of
the $\gamma\delta$-chains of Γ' is separated from Q by an $\alpha\beta$-chain, and so has
no residue in Γ. There are three others. Two of them have residues
consisting of a single Kempe chain of Γ each. The vertex-sets are
$\{G\}$ and $\{I, J\}$. The residue of the third consists of two Kempe
chains of Γ. The vertex sets are $\{B\}$ and $\{D\}$. The arrows marked
on some of the edges are intended to clarify part of the following
proof:

THEOREM 8.3: *Let U be a near-triangulation bounded by Q, let Γ
be a 4-colouring of Q and let Γ' be a U-extension of Γ. Let Z be the
family of residues in Γ of the $\alpha\beta$-chains and $\gamma\delta$-chains of Γ'. Then Z
is an admissible clustering of Γ with respect to $(\alpha\beta, \gamma\delta)$.*

Proof: That Z is a clustering of Γ follows from the fact that
each vertex of Q belongs to exactly one $\alpha\beta$-chain or $\gamma\delta$-chain of
Γ'.

By 4.1, Z satisfies Condition (i) for an admissible clustering.

To prove Condition (ii) for Z let C_1 be any cluster in Z and let
L be an arc in Q, with ends B and D say, such that no vertex of L
is included in C_1 but B and D are adjacent in Q to vertices A and
E respectively (not necessarily distinct) that are included in C_1.

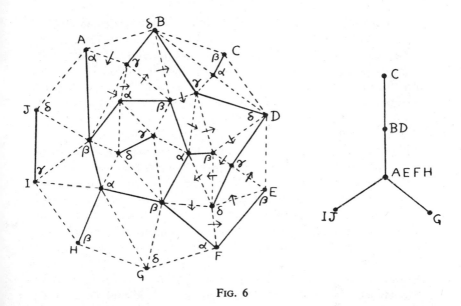

FIG. 6

Let C_2 be the member of Z that includes B. We have to show that C_2 also includes D.

Let C_1 and C_2 be residues of Kempe chains K_1 and K_2 of Γ' respectively. We may suppose K_1 to belong to $\{\alpha, \beta\}$ and K_2 to $\{\gamma, \delta\}$.

Let us define a *crossing edge* of U as an edge with one end in K_1 and one in K_2. Let a *crossing triangle* of U be a face, other than the one bounded by Q, that is incident with a vertex of K_1 and a vertex of K_2. We assume in this definition that only one face of U can be bounded by Q. If two are then Q is a triangle identical with U, and the required result is trivial. Evidently the following rules hold:

(iii) *If T is a crossing triangle, then each vertex of T is in K_1 or K_2.*

(iv) *Each crossing triangle is incident with exactly two crossing edges.*

Consider the crossing edge AB of Q. It is incident with a single crossing triangle T_1, and this has a unique second crossing edge t_1. If t_1 is not an edge of Q it is incident with a second crossing triangle T_2, and this has a second crossing edge t_2. If t_2 is not an edge of Q it is incident with a second crossing triangle T_3, and so on. Such a sequence is indicated in Figure 6. T_1 is the triangle incident with AB. An arrow leads from this triangle into T_2, another arrow from T_2 into T_3, and so on.

Suppose some triangle is repeated in the sequence T_1, T_2, ... Let us say $T_i = T_j$, where $i < j$ and j has the least value consistent with this condition. Now t_{j-1} is either t_{i-1} or t_i. If $t_{j-1} = t_{i-1}$, then $T_{j-1} = T_{i-1}$, contrary to the choice of j. Hence $t_{j-1} = t_i$, $T_{j-1} = T_{i+1}$ and, by the choice of j, $j - 1 = i + 1$. But now $t_{i+1} = t_{j-1} = t_i$, which is impossible.

We conclude that in fact no triangle is repeated in the sequence. The sequence therefore terminates with a triangle T_m such that t_m is an edge of Q.

The crossing edge t_m has one end X_1 in K_1 and one end X_2 in K_2. Hence X_1 and X_2 are included in the clusters C_1 and C_2 respectively. Following along L from B we cannot reach X_1 before reaching E. Applying (i) to A, E, B and X_2 we find that X_2 cannot lie beyond E. Hence $X_2 = D$, C_2 includes D, and the proof is complete.

We refer to Z as the admissible clustering of Γ, with respect to $(\alpha\beta, \gamma\delta)$, *induced by* Γ'. Similarly the corresponding chromodendron χ is the *chromodendron of* Γ, with respect to $(\alpha\beta, \gamma\delta)$, *induced by* Γ'. An example of such a chromodendron is shown in Figure 6.

We need the following converse to 8.3:

THEOREM 8.4: *Let Γ be a 4-colouring of a circuit Q, and let χ be a chromodendron of Γ with respect to $(\alpha\beta, \gamma\delta)$. Then we can construct a near-triangulation U bounded by Q, and a U-extension Γ' of Γ such that Γ' induces χ by way of the residues of its $\alpha\beta$-chains and $\gamma\delta$-chains.*

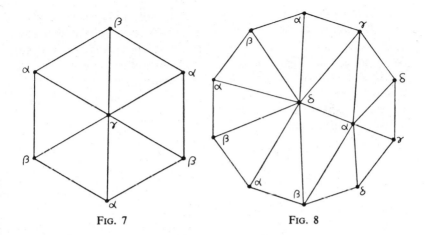

Fig. 7   Fig. 8

Proof: Let Z be the admissible clustering of Γ corresponding to χ.

Let n be the number of Kempe chains of Γ belonging to $\{\alpha, \beta\}$ or $\{\gamma, \delta\}$. For the construction of U in the cases $n = 1$ and $n = 2$ see Figures 7 and 8 respectively.

Let us assume as an inductive hypothesis that the theorem is true whenever n is less than some integer $q \geqslant 3$, and let us consider the case $n = q$. We can suppose q to be even, since Kempe chains of Γ belonging to $\{\alpha, \beta\}$ and $\{\gamma, \delta\}$ occur alternately in Q.

Evidently we can find vertices A and B of Q, each coloured α or β, and vertices C and D of Q, each coloured γ or δ, such that A and B separate C and D in Q. It may happen that A and B are included in the same cluster of Z. But suppose not. Then there is an arc M in Q that contains B but no vertex included in the same cluster as A, and has the maximum number of vertices consistent with this condition. We can take C and D to be the ends of M since these ends will not be coloured α or β. Then C and D are each adjacent to a vertex included in the same cluster as A, since otherwise M could be extended. It follows from Condition (ii) that C and D are included in a common cluster of Z.

Adjusting the notation we can assert that Q has two vertices A and B, each coloured α or β, belonging to different $\alpha\beta$-chains J and K respectively of Q, but included in the same cluster W of Z. We join A and B by an arc L inside Q as the first step in the construction of U. If A and B have different colours, L consists of a single edge. Otherwise L consists of two edges, and their common end is given a colour α or β different from the colour of A and B. (See Figure 9.)

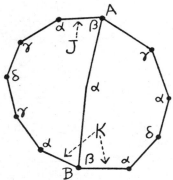

FIG. 9

Now Q is the union of two arcs L_1 and L_2 with common ends A and B but otherwise disjoint. We write Q_1 and Q_2 for the circuits $L \cup L_1$ and $L \cup L_2$ respectively. The colours already assigned determine 4-colourings Γ_1 and Γ_2 of Q_1 and Q_2 respectively.

Let j be 1 or 2. The Kempe chains of Γ_j belonging to $\{\alpha, \beta\}$ or $\{\gamma, \delta\}$ are the correspondingly coloured Kempe chains of Γ contained in L_j, together with an $\alpha\beta$-chain H_j which is the union of L with $J \cap L_j$ and $K \cap L_j$. Let us write n_j for the number of $\alpha\beta$-chains and $\gamma\delta$-chains of Γ_j. Then evidently $n_j < q$. By Condition (i) each cluster of Z includes vertices of only one of the arcs L_1 and L_2, except for the cluster W. Hence Γ_j has a clustering Z_j consisting of those clusters of Z that include only vertices of L_j, together with a cluster W_j defined as follows. The members of W_j are those members of W, other than J and K, that are contained in L_j, together with H_j. It can now be shown that Z_j is an admissible clustering of Γ_j. The proof is quite straightforward, and to save space we leave it to the reader.

Since $n_j < q$ it follows by the inductive hypothesis that we can triangulate the inside of Q_j to form a near-triangulation U_j bounded by Q_j, arranging that Γ_j has a U_j-extension Γ_j' the residues of whose $\alpha\beta$-chains and $\gamma\delta$-chains are the clusters of Z_j.

Combining U_1 and U_2 we obtain a near-triangulation U bounded by Q, and we can combine Γ_1' and Γ_2' to obtain a U-extension Γ' of Γ. The $\alpha\beta$-chains and $\gamma\delta$-chains of Γ' are those of Γ_1' and Γ_2', except that those containing L are replaced by their union, an $\alpha\beta$-chain of Γ' whose residue is the cluster W. Thus Z is induced by Γ'.

The theorem is now established for the case $n = q$. It follows in general by induction.

The next theorem reduces to finiteness the problem of the dominance of a given set of 4-colourings of a circuit.

THEOREM 8.5: *Let Γ be a 4-colouring of a circuit Q, and let S be a set of 4-colourings of Q. Then Γ is simply immersible in S if and only if the following condition holds: There exists a colour-partition Π such that, for each chromodendron χ of Γ with respect to Π, Γ is χ-transformable into a member of S.*

Proof: Suppose Γ is simply immersible in S. Then there is a colour-partition Π with the following property: if U is any near-triangulation bounded by Q and if Γ is U-extensible, then there is a U-extension Γ' of Γ that can be transformed into a U-extension of a member of S by a succession of Kempe interchanges with respect to Π. By 8.4 we can choose U and Γ' so that Γ' induces χ, where χ is an arbitrary chromodendron of Γ with respect to Π. In Q the succession of Kempe interchanges reduces to a succession of simple χ-transformations, taking Γ into a member of S. Thus the stated condition holds.

Conversely suppose the condition to hold. Let U be any near-triangulation bounded by Q such that Γ has a U-extension Γ'. Let Γ' induce the clustering Z of Γ, corresponding to a chromodendron χ. There exists a succession of simple χ-transformations changing Γ into a member of S. Each affects a single cluster of Z, and each can be effected by a Kempe interchange applied to the Kempe chain of Γ' having that cluster as its residue. In this way we construct a succession of Kempe interchanges in U, with respect to Π, transforming Γ' into a U-extension of a member of S. Thus Γ is simply immersible in S.

Let us consider how to test for dominance a set S of 4-colourings of a circuit Q. Consider any 4-colouring Γ of Q not in S. Let $J_1(\Gamma)$, $J_2(\Gamma)$ and $J_3(\Gamma)$ be the sets of all chromodendra of Γ with respect to $(\alpha\beta, \gamma\delta)$, $(\alpha\gamma, \beta\delta)$ and $(\alpha\delta, \beta\gamma)$ respectively. We begin by determining these three sets. Then for each $\chi \in J_1(\Gamma)$ we determine all the 4-colourings Γ'' of Q such that χ admits a transformation of Γ into Γ''. It may happen that for each such χ one of the 4-colourings Γ'' is in S. If so we have $\Gamma \in f(S)$, by 8.5. If not we repeat the procedure with $J_2(\Gamma)$, and then if necessary with $J_3(\Gamma)$. If for some i each chromodendron of $J_i(\Gamma)$ admits a transformation of Γ into a member of S, then $\Gamma \in f(S)$, but otherwise Γ is not in $f(S)$. Repeating the whole procedure for each 4-colouring Γ not in S we complete the determination of the set $f(S)$. Then we can find $f^2(S)$ in the same way, and so on. Eventually the process must terminate with a set $f^k(S)$ identical with $f^{k+1}(S)$. (A set T such that $f(T) = T$ will be called *closed*. Perhaps the term "*D*-closed" would be better; but we shall not make use of other kinds of reducibility in this connection.) If $f^k(S)$ includes all the 4-colourings of Q, then S is dominant; otherwise it is not.

This is the procedure applied by Heesch to the set S of V-extensible 4-colourings of Q to determine if a pair (V, Q) is D-reducible, that is if S is dominant. Let us call it *Heesch's algorithm*.

It can be shown that if (V, Q) is D-reducible, then (V, Q) is K-reducible. The proof uses the fact that in the second paragraph of the proof of 8.5 Γ' can be any U-extension of Γ. We leave it to the reader to bridge the gap due to the fact that we have defined

K-reducibility in terms of general planar graphs, and D-reducibility only in terms of near-triangulations.

Perhaps we should take note here of some other kinds of reducibility defined by Heesch [3]. If V is a near-triangulation bounded by Q, then (V, Q) is said to be *A-reducible* if the following condition holds: there is a near-triangulation W bounded by Q, having fewer vertices than V, and such that every W-extensible 4-colouring of Q is in the set S of V-extensible 4-colourings of Q. To prove that (V, Q) is then reducible consider a supposedly minimal graph G separated into V and another near triangulation U by the circuit Q. The union of U and W, we deduce, is 4-colourable. Hence there is a 4-colouring of Q that is U-extensible and W-extensible, and therefore V-extensible. This implies that G has a 4-colouring and we have a contradiction.

B-reducibility is similarly defined except that it requires the W-extensible 4-colourings of Q to be simply immersible in S, not necessarily all in S. The definition of C-reducibility replaces "simply immersible" by "crudely immersible." In each case the proof of reducibility is a straightforward generalization of the one we have sketched in the case of A-reducibility. Heesch gives examples from the literature of these three kinds of reducibility.

In the theory of D-reducibility we try to make as much progress as possible without considering the effect of a Kempe interchange on the Kempe chains of other colour-partitions. Perhaps the Four Colour Problem can be settled in this way, but perhaps the drastic simplification rejects essential information. The authors would feel much more confident in the power of the theory if it could be used to prove that the Birkhoff number exceeds some reasonable number, —say 40, or even 20.*

9. A CONSTRUCTION FOR D-IRREDUCIBLE PAIRS.

We now give a method for constructing D-irreducible pairs (V, Q). Of course an attempt to prove the Four Colour Conjecture

---

*Most of the reductions used to assign a lower bound to the Birkhoff number have now been identified as D-reductions.

uses, rather, D-reducible pairs; thus the present construction puts difficulties in the way of the method. It shows directions in which the method need not be pursued.

The authors feel tempted to construct an abstract theory of dominant sets. They have defined a *closed* set of 4-colourings of a circuit Q towards the end of Section 8. The definition amounts to saying that a set T is closed if no member of its complement is simply immersible in T. A set R can be called *open* if its complement is closed, that is if no member of R is simply immersible in the complement of R. Here is an example of a theorem from the abstract theory of dominant sets.

THEOREM 9.1: *Let J be the set of all 4-colourings of a circuit Q. Then each non-null open subset of J meets every dominant subset of J.*

Proof: Let S be a dominant subset of J, and T a non-null open subset of J. Assume that their intersection is null. There is a least integer k such that $f^k(S)$ meets T. Moreover $k > 0$. Choose $\Gamma \in T \cap f^k(S)$. Then Γ is simply immersible in $f^{k-1}(S)$ and therefore it is simply immersible in the complement of T. This is contrary to the hypothesis that T is open.

Let us define a *pentatriangulation* as a plane graph in which each face is either a triangle or a pentagon. A *near-pentatriangulation* is a plane graph in which each face with at most one exception is a triangle or a pentagon. When we say that a near-pentatriangulation is bounded or face-bounded by a circuit Q we shall imply that Q is the face-boundary of the exceptional face, if there is one.

A *P-colouring* of a pentatriangulation G is a 4-colouring of G such that all four colours appear on the boundary of each pentagonal face. A P-colouring of a near-pentatriangulation G bounded by a circuit Q is defined in the same way, with the understanding that the four-colour condition may be relaxed for Q if Q is a pentagon and not the whole of G.

THEOREM 9.2: *Let U be a near-pentatriangulation bounded by a* *circuit Q. Let T be the set of all* 4-*colourings of Q having U-extensions that are P-colourings of U. Then T is open.*

Proof: Let Γ be a 4-colouring of Q having a U-extension Γ' that is a P-colouring of U. We convert U into a near-triangulation bounded by Q as follows. Consider any pentagonal face P_j of U satisfying the four-colour condition. Let its face-boundary be Q_j. In Γ' one colour, say α, is repeated in Q_j. (See Figure 5.) Call the vertices coloured α "special", and call the vertex between them the "apex". In each face P_j we join the apex to the two non-adjacent vertices of Q_j. We observe that Γ' is preserved as a 4-colouring of the resulting near-triangulation U'. We note also that no succession of Kempe interchanges in Γ' and U', provided they are all with respect to the same colour-partition, can transform Γ' into a 4-colouring of U' in which only three colours appear on one of the circuits Q_j. This means that Γ is not simply immersible in the complement of T. We deduce that T is open.

THEOREM 9.3: *Let a circuit Q separate a pentatriangulation G into a near-pentatriangulation U and a near-triangulation V, both bounded by Q. Suppose U to have a P-colouring, but G to have no P-colouring. Then (V, Q) is D-irreducible.*

Proof: Assume the contrary. Let S be the set of V-extensible 4-colourings of Q, and T the set of 4-colourings of Q having U-extensions that are P-colourings of U. Then S is dominant. T is non-null by hypothesis, and open by 9.2. Hence S and T have a common member Γ, by 9.1. Combining a V-extension and a suitable U-extension of Γ we can obtain a P-colouring of G, contrary to hypothesis.

We go on to describe a construction due to Shimamoto.
Suppose there is a critical graph containing a 5-wheel, any minimal graph for example. Dropping out the hub and spokes of this wheel gives a plane graph G^* that we call a "chromatic

obstacle". It is bounded by the pentagonal rim of the wheel. It is 4-colourable, but all four colours must appear on the rim.

In the construction as described by Shimamoto at the time of the rumours new critical graphs are built up from smaller ones. It is found that a critical graph obtained in this way must have the following structure. There is a pentatriangulation X with no P-colouring, and the critical graph is obtained from X by filling each face with a chromatic obstacle. It is possible therefore to describe the construction as one in which new pentatriangulations without P-colourings are obtained from old ones. We prefer to describe it in this way and so to avoid any use of such hypothetical figures as critical graphs and chromatic obstacles.

The principal step in the construction is based on the following fact. Suppose Q bounds U and V, the whole forms G, and the conditions of 9.3 hold. Let A, X, B be three consecutive vertices of Q. Let us cut along the arc AXB and open it out into a quadrilateral $AXBX'$, with X still on the boundary of V. Finally let us fill the new quadrilateral with three triangles and a pentagon, as shown in Figure 10. Let us now replace the arc AXB in Q by the arc AY_1Y_2B, to form a circuit Q'. There is a near-triangulation V' defined by the faces of G in V together with the three new triangles. This is separated by Q; from a near-pentatriangulation U' defined by the faces of G in U together with the new pentagon AY_1Y_2BX'. The new pentatriangulation that is the union of U' and V' we denote by G'. It is appropriate to call this operation Shimamoto's First Construction. The construction is reminiscent of one used by Hajós for critical graphs [5, Section 11.4].

The conditions of 9.3 hold with Q', U' and V' replacing Q, U and V respectively. We know U has a P-colouring, and we can extend this as a P-colouring of U' by assigning appropriate colours to the new vertices Y_1 and Y_2, both divalent in U'. If G' has a P-colouring it is clear that X and X' must have the same colour in it, and so we can derive a P-colouring of G, which is impossible. We have derived a new D-irreducible pair (V', Q'). We can now drop the primes and repeat the construction.

As our starting point we can take G to be the 5-wheel and Q its rim. U is the circuit Q, and V is the 5-wheel itself. U has two pentagonal faces, both bounded by Q. It is thus a near penta-

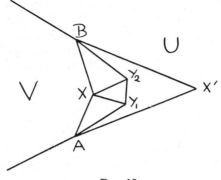

Fᴉɢ. 10

triangulation bounded by Q. In a P-colouring of U each face must have all four colours in its boundary. Evidently the conditions of 9.3 hold. Thus 9.3 shows that the 5-wheel is D-irreducible, as we have seen already in 7.2, Corollary.

Figure 11 shows the results of some successive applications of Shimamoto's Construction to the 5-wheel. In each diagram the circuit Q is shown by arrows. An asterisk indicates the vertex X that is to be split.

For the final step we start with the last diagram of Figure 11 and omit the broken edges. We take two copies of the resulting graph and identify their arcs AKL. We then introduce new vertices R, A_1' and A_2', and join from them as shown in Figure 12.

This is an example of Shimamoto's Second Construction.

Let us show that the graph G_{12} of Figure 12 satisfies the conditions of 9.3 with Q as the circuit Q_{12} indicated by the arrows, V as the near-triangulation, which we shall also call H, inside Q_{12}, and U as the near-pentatriangulation U_{12} outside Q_{12}.

We prove that G_{12} has no P-colouring as follows: If it did, then using Σ shows that A has the same colour as A_1' or A_2'; say $\text{col}(A) = \text{col}(A_1')$. Using Σ_B now shows that $\text{col}(B) = \text{col}(B')$. Using Σ_C and Σ_D in turn shows that $\text{col}(C) = \text{col}(C')$ and $\text{col}(D) = \text{col}(D')$. Noting that K is joined to all of A, B, C, D and L now shows that Σ' uses only three colours, a contradiction.

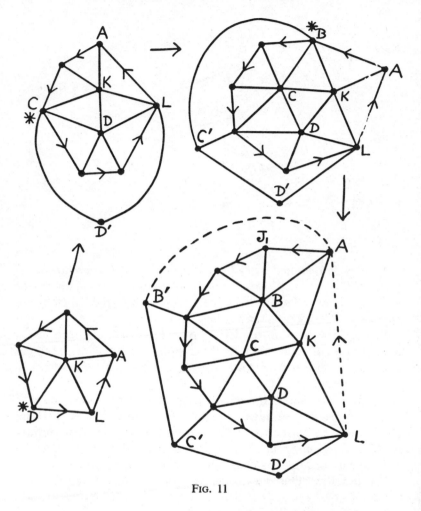

FIG. 11

It remains to construct a *P*-colouring for U_{12}, shown completely in Figure 12. Ignoring R and the edge $A_1'A_2'$ we find that this graph consists of two isomorphic parts with only the vertex L in common. From a *P*-colouring of the graph U of the last diagram of Figure 11 we can, for each part, derive a 4-colouring that obeys the four colour rule for pentagonal faces of G_{12} bounded by circuits in the part considered. After a permutation in one of the

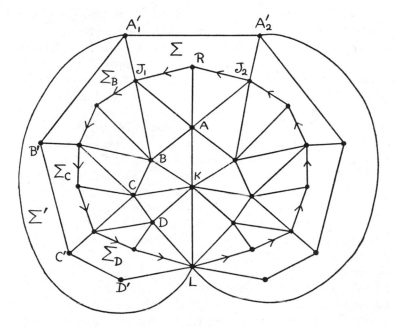

FIG. 12

two parts we can combine the two 4-colourings to obtain a 4-colouring of U_{12}, less R, such that A_1' and A_2' have different colours and at least three colours occur among the four vertices A_1', A_2', J_1 and J_2. We can now colour R so as to complete a P-colouring of U_{12}.

We can now apply 9.3 to obtain

THEOREM 9.4: *The near-triangulation H is D-irreducible.*

10. CONCLUDING REMARKS.

The configuration H of 9.4 is the one mentioned in Section 1. At first it was thought to have been proved D-reducible by a computer programmed to apply Heesch's Algorithm. In the original form of Shimamoto's Construction the pentagonal faces of G_{12} were supposed to be filled with chromatic obstacles. Then U_{12} was a near-triangulation and it could be proved to have a 4-colouring. Arguments resembling those of Section 9 led to a contradiction corresponding to our 9.4. It seemed that this could only be resolved by supposing that U_{12} was impossible, i.e., that no chromatic obstacles existed and the Four Colour Conjecture was true. Now it is clear that Shimamoto had discovered not a proof of the Conjecture but a construction for D-irreducible configurations.

To the present authors the supposed D-reducibility of H meant that any 4-colouring of U_{12} could be converted into one extendable to all G_{12} by crude chaining applied to U_{12}. Somehow in the course of this chaining the four colours on the boundary of one of the supposed chromatic obstacles would reduce to three. By isolating the effect of the chaining on this one chromatic obstacle we should obtain a proof by crude chaining of the reducibility of the 5-wheel. Yet it seemed clear that our 7.3, a well-known result, was the best that could be expected along this line. It was to rigorize this objection that we introduced P-colourings and worked out their theory as given in Section 9.

In this report on the present state of the Four Colour Problem there is little for which we claim originality, apart from the recognition of the true meaning of Shimamoto's Construction. We have tried to clarify the theory for ourselves, and we dare to hope that we may thereby have clarified it for others.

It now seems to us that the next step in the theory of D-reducibility should be an attempt to classify the minimal dominant sets for the smaller circuits. To test (V, Q) for D-reducibility we would then determine the set S of V-extensible 4-colourings of Q and check it against a list of dominant sets of Q to see if it contained one of them. Should this method prove feasible it might eliminate much repetitive work.*

---

*In the original version of this paper the authors proposed a conjecture here to the effect that any two non-null open sets of 4-colourings of a circuit must intersect. It has been pointed out, however, that simple counter-examples exist, one being provided by the sets of full and non-full 4-colourings of the pentagon.

REFERENCES

1. Birkhoff, G. D., "The Reducibility of Maps," *Amer. J. Math.*, **35** (1913), 115–128.

2. Heawood, P. J., "Map-colour Theorem," *Quart. J. Math. Oxford*, Ser. **24** (1890), 322–338.

3. Heesch, H., *Untersuchungen zum Vierfarbenproblem*, Hochschulskripten 810/810a/810b, Mannheim 1969.

4. Kempe, A. B., "On the Geographical Problem of the Four Colours," *Amer. J. Math.*, **2** (1879), 193–200.

5. Ore, O., *The Four-color Problem*, Academic Press, New York, 1967.

6. Ore, O., and Stemple, J., "Numerical Calculations on the Four-color Problem," *J. Combinatorial Theory*, **8** (1970), 65–78.

7. Saaty, T. L., "Thirteen Colorful Variations on Guthrie's Four-color Conjecture," *Amer. Math. Monthly*, **79** (1972), 2–43.

8. Tutte, W. T., "On the Four Colour Conjecture," *Proc. London Math. Soc.*, **50** (1948), 137–149.

9. Winn, C. E., "On the Minimum Number of Polygons in an Irreducible Map," *Amer. J. Math.*, **62** (1940), 406–416.

AUTHOR INDEX

This index covers MAA Studies Volume 11 (Studies in Graph Theory, Part I, pages 1 to 199) and MAA Studies Volume 12 (Studies in Graph Theory, Part II, pages 201–413).

SUBJECT INDEX

This index covers MAA Studies Volume 11 (Studies in Graph Theory, Part I, pages 1 to 199) and MAA Studies Volume 12 (Studies in Graph Theory, Part II, pages 201–413).

15